AF753505

ADVANCES IN ENERGY RESEARCH

VOLUME 31

ADVANCES IN ENERGY RESEARCH

Additional books and e-books in this series can be found on Nova's website under the Series tab.

ADVANCES IN ENERGY RESEARCH

ADVANCES IN ENERGY RESEARCH

VOLUME 31

MORENA J. ACOSTA
EDITOR

nova
science publishers
New York

Copyright © 2019 by Nova Science Publishers, Inc.

All rights reserved. No part of this book may be reproduced, stored in a retrieval system or transmitted in any form or by any means: electronic, electrostatic, magnetic, tape, mechanical photocopying, recording or otherwise without the written permission of the Publisher.

We have partnered with Copyright Clearance Center to make it easy for you to obtain permissions to reuse content from this publication. Simply navigate to this publication's page on Nova's website and locate the "Get Permission" button below the title description. This button is linked directly to the title's permission page on copyright.com. Alternatively, you can visit copyright.com and search by title, ISBN, or ISSN.

For further questions about using the service on copyright.com, please contact:
Copyright Clearance Center
Phone: +1-(978) 750-8400 Fax: +1-(978) 750-4470 E-mail: info@copyright.com.

NOTICE TO THE READER

The Publisher has taken reasonable care in the preparation of this book, but makes no expressed or implied warranty of any kind and assumes no responsibility for any errors or omissions. No liability is assumed for incidental or consequential damages in connection with or arising out of information contained in this book. The Publisher shall not be liable for any special, consequential, or exemplary damages resulting, in whole or in part, from the readers' use of, or reliance upon, this material. Any parts of this book based on government reports are so indicated and copyright is claimed for those parts to the extent applicable to compilations of such works.

Independent verification should be sought for any data, advice or recommendations contained in this book. In addition, no responsibility is assumed by the Publisher for any injury and/or damage to persons or property arising from any methods, products, instructions, ideas or otherwise contained in this publication.

This publication is designed to provide accurate and authoritative information with regard to the subject matter covered herein. It is sold with the clear understanding that the Publisher is not engaged in rendering legal or any other professional services. If legal or any other expert assistance is required, the services of a competent person should be sought. FROM A DECLARATION OF PARTICIPANTS JOINTLY ADOPTED BY A COMMITTEE OF THE AMERICAN BAR ASSOCIATION AND A COMMITTEE OF PUBLISHERS.

Additional color graphics may be available in the e-book version of this book.

Library of Congress Cataloging-in-Publication Data

ISBN: 978-1-53616-271-4
ISSN: 2157-1562

Published by Nova Science Publishers, Inc. † New York

CONTENTS

PREFACE

In the past, fossil fuels have been considered a reliable source of energy by many countries. Despite the current measures to reduce the use of fossil fuels for electricity generation, the use of this type of energy source will continue for the foreseeable future. Advances in Energy Research. Volume 31 examines how, if this situation does not change in the future, then all types of fossil fuels, particularly oil, will be exhausted before the end of the current century.

The use of microholes may potentially enhance the applications of fossil fuels in various energy fields, especially in thermal power plants. Therefore, the morphology variation of micropunch and microhole during the punching with WC/Co micropunches is investigated in this compilation through scanning electron microscopy, energy dispersive spectrometer and confocal laser studies.

Following this, the authors provide a review on experimental and modeling studies performed with direct fuel cells, starting with the fundamentals and main drawbacks of these systems, followed by an exhaustive review on the different studies performed regarding the effects of operating conditions and design parameters.

In the penultimate chapter, a new method for determining the cohesion potential energy in solids is described, and fundamental factors such as internal energy, entropy, temperature and time are examined, in particular the connection of entropy and time to potential energy. The method concerns molecular solids and rests upon measurable thermodynamic quantities: heat of sublimation, heat capacity, and entropy as derived from the latter.

The closing chapter provides an overview of three methods (retarded osmosis method, forward osmosis-electrokinetic method, and diffusio-osmotic method) for sustainable power generation from the salinity gradient energy by using osmosis-based processes.

Chapter 1 - Fossil fuels were considered in the past by many countries as a reliable source of energy. Despite the measures already adopted by many countries to reduce the use of fossil fuels for electricity generation, the use of this type of energy source will continue to play an essential role within the energy mix of these countries at least until 2050. But fossil fuel reserves are limited, and currently, it is consumed at a much higher rate than what is produced. If this situation does not change in the future, then all types of fossil fuels, particularly oil, will be exhausted probably before the end of the current century. Despite the multiple uses of fossil fuels in different economic sectors, many governments and organizations around the world seek to change the current pattern of CO_2 emissions and achieve the goal set in the Paris Agreement to keep the global temperature rise below 2°C. To accomplish this goal, a significant advance in innovation and development of the different types of available renewable energies should be achieved, in addition to an increase in energy efficiency, and the use of more efficiency accumulation systems and smart grids. In the case of coal, the capture and storage of carbon, especially for natural gas-fired power plants, should continue to be part of the energy landscape in several countries due to its flexible availability for electricity generation and its less negative impact on the environment and the population. However, fossil fuels have experienced a renaissance in recent years, thanks to modern prospecting techniques, which have made it possible to find new deposits that were unknown until a few decades ago. Although most of the fossil fuels when burned produce a negative impact on the environment and the population, it is expected that the use of this type of energy sources, particularly the use of natural gas for electricity generation, will continue to be an essential component of the energy mix of several countries all over the world. On the other hand, the use of oil and coal for electricity generation it is expected to decrease significantly by 2050, but they will continue to be an essential component of the energy mix in several countries at least until that year.

Chapter 2 - The use of microholes is a potential approach to enhance fossil fuels efficient applications in the various energy fields, especially for thermal power plants. Ascribed to eco environmental competition, micropunching is extremely suitable for the mass production of micro features with friendly eco effects. Therefore, the morphology variation of micropunch and microhole during the punching with WC/Co micropunches was investigated by scanning electron microscopy (SEM), energy dispersive spectrometer (EDS) and confocal laser. The results reveal that due to the optimal joint-contribution of WC and Co, the formed microholes satisfy with the practical requirements in the quasistable period. Moreover, the serious wear of micropunch occurs with the wear loss both of Co and WC when the punching number exceeds 1525. With the punching numbers further increment, the dominant factors of the wear loss would mainly rely on the easily peeled off WC due to the serious loss of Co. In addition, the microholes can be adequately processed after about 30 min by natural sand grains. The quality of the hole would decrease with further increase in processing time and sand accumulation becomes severe.

Chapter 3 - Fuel cells are being extensively studied due to their potential as an alternative energy supply technology for a wide range of applications. Among the different types of fuel cells available, direct fuel cells (DFCs) are promising ones to provide energy for electronic devices, since they offer high energy densities, long runtime, instant recharging and use liquids as fuel. Ethanol emerges as a very attractive fuel when compared with the other alcohols commonly used in DFCs, since it is less toxic, has higher energy density and can be produced from renewable sources. Based on that, direct ethanol fuel cells (DEFCs) emerge as a good choice for producing sustainable energy and have received great attention, in the last years, both in fundamental and technological aspects. However, they are still far from attaining acceptable levels of power output, since their performance is affected by the slow electrochemical ethanol oxidation and water and ethanol crossover. This work provides a review on the experimental and modeling studies performed with DEFCs, towards its optimization. Therefore, this work starts with the fundamentals and main drawbacks of these systems, followed by an exhaustive review on the different studies performed regarding the effect of operating conditions (ethanol concentration, air pressure, anode and cathode flow rate and temperature) and design parameters (membrane electrode assembly characteristics/ materials and flow field effect) on its power output and performance. As mathematical models are a fundamental tool for the design and optimization of fuel cells systems, this work also includes a section devoted to the modeling studies accomplished for DEFCs. Towards the use of DEFCs in portable electronic devices, different studies performed with passive direct ethanol fuel cells are presented at the end of this work.

Chapter 4 - In recent years, the demand for efficient, renewable and sustainable energy sources has been increasing in order to obtain a more diverse energy production. Direct alcohol fuel cells (DAFCs) appear as a very promising power solution for portable devices, since they offer high energy densities and longer runtime than batteries, instant recharging, use a liquid fuel (alcohol), operate at room temperature and can rely on a passive flow of reactants and products, allowing to operate the cell without any additional power consumption. Therefore, much work has been done towards the optimization of passive DAFC systems, where the fuel and the oxidant are feed in a passive mode, and some progress towards its commercialization has already been achieved. Despite that, these systems still have costs above the desirable and a lower lifetime and power output. The passive DAFCs can be supplied with different alcohols, but the most used and studied ones are methanol and ethanol. Methanol has the highest energy to carbon ratio of any other alcohol, but is toxic for humans and is non-renewable. Ethanol emerges as an alternative and promising fuel due to its non-toxicity, natural availability and renewability. Nevertheless, the performance of direct ethanol fuel cells (DEFCs) is lower than the DMFCs (direct methanol fuel cells), due to the sluggish ethanol electrochemical oxidation, since it is very difficult to cleavage the carbon-carbon bond presented on the ethanol structure. A novel idea to achieve higher performances with passive DAFCs is to use a mixture of methanol and ethanol, since it is expected to obtain the best characteristics of each alcohol. In order to achieve that, the present chapter presents a study with a passive DAFC fed with a mixture of methanol and ethanol with different concentrations and volume ratios, towards its optimization. The experimental results were analyzed under the light of the polarization and electrochemical impedance spectroscopy (EIS) measurements. The EIS technique was used to identify and quantify the different losses that negatively affect the passive DAFC performance, pointing out possible solutions to minimize them.

Chapter 5 - Potential energy in materials relates to the development of interaction energy of electrostatic or electronic character between the particles of the chemical entity constituting a material. The interaction energy develops when the particles are assembled to form the material starting with the particles ideally separated like a gas. The forces of cohesion between atoms, ions or molecules forming materials comprise weak interactions of dipole character, and weak and strong chemical bonding, depending on the type of material. Potential energy of cohesion in solids correlates with the vibrations of the lattice of the solid at the outside of the constituent chemical entity. The dissociation potential energy of the single oscillator characteristic to lattice vibrations equals the potential energy of the lattice. A new method to find out the cohesion potential energy in solids is described, and fundamental factors internal energy, entropy, temperature and time are examined as connected to potential energy in processes, in particular the connection of entropy and time to potential energy. The method concerns molecular solids and rests upon measurable thermodynamic quantities: heat of sublimation, heat capacity, and entropy as derived from the latter. Valuation of cohesion potential energy by the heat of sublimation requires knowledge of the kinetic energy for molecules both in the solid and the gas, the cohesion energy being the second component of the overall energy in the solid. The vibrational energy for the solid can be known from the heat capacity using instrumentally Debye's theory yet a molecular solid is not monoatomic as prescribed in the theory. The mathematical function for entropy in this theory fits sufficiently better to entropy in the poliatomic molecular solid than the function for heat capacity to the heat capacity of the solid. Accuracy in determining vibrational energy involves the use of the entropy function. Vibrational energy for the gas molecules is difficultly obtainable with sufficient accuracy by heat capacity with simple harmonic oscillator model and can be attained by entropy and Debye entropy function as applied to the gas. Besides cohesion potential energy the method allows for determination of absolute entropies of solids and cognate melts and vapor phases using a restricted set of heat capacity data. In molecular liquids Debye function for entropy is suitable to valuate jointly vibrational and rotational entropies. Cohesion potential energy is embodied into the translation entropy of molecules in a liquid, this feature of potential energy being connected to the relationship of entropy to time. The method yields the cohesion potential energy for a liquid added up to the bringing of internal electronic energy, when this proceeds by facile diaelectrical mobility in the liquid. The efficacy of potential energy in a spontaneous transformation or process with minimal thermic and entropic dispensing relies on the speed of the transition.

Chapter 6 - Salinity gradient energy is referred to as the chemical potential difference between two solutions with different concentrations. It commonly occurs at river mouths where a river flows into a sea/ocean and it has an equivalent hydraulic head as high as 270 m, which is also termed as a 'silent waterfall'. Deterioration in environment and depletion of conventional energy resources arise the global need in finding alternative energy sources. The salinity gradient energy, as the second largest marine-based energy source and one of the most potential energy resources, has attracted an increasing attention over the past decade. Osmosis is considered as a potential way to harvest this kind of salinity gradient energy. Extensive research works have been conducted to develop green and environment-friendly power generation methods based on osmosis-related mechanisms, and they include pressure-retarded osmosis (PRO) method, forward osmosis-electrokinetic (FO-EK) method, and diffusio-osmotic (DO) method. This chapter provides an overview of these three methods for sustainable power generation from the salinity gradient energy by using osmosis based processes.

In: Advances in Energy Research. Volume 31 ISBN: 978-1-53616-271-4
Editor: Morena J. Acosta © 2019 Nova Science Publishers, Inc.

Chapter 1

FOSSIL FUELS: CURRENT SITUATION AND PERSPECTIVES OF ITS USE AT WORLD LEVEL

Jorge Morales Pedraza*
Morales Project Consulting, Vienna, Austria

ABSTRACT

Fossil fuels were considered in the past by many countries as a reliable source of energy. Despite the measures already adopted by many countries to reduce the use of fossil fuels for electricity generation, the use of this type of energy source will continue to play an essential role within the energy mix of these countries at least until 2050. But fossil fuel reserves are limited, and currently, it is consumed at a much higher rate than what is produced. If this situation does not change in the future, then all types of fossil fuels, particularly oil, will be exhausted probably before the end of the current century.

Despite the multiple uses of fossil fuels in different economic sectors, many governments and organizations around the world seek to change the current pattern of CO_2 emissions and achieve the goal set in the Paris

* Corresponding Author's E-mail: jmorales47@hotmail.com.

Agreement to keep the global temperature rise below 2°C. To accomplish this goal, a significant advance in innovation and development of the different types of available renewable energies should be achieved, in addition to an increase in energy efficiency, and the use of more efficiency accumulation systems and smart grids. In the case of coal, the capture and storage of carbon, especially for natural gas-fired power plants, should continue to be part of the energy landscape in several countries due to its flexible availability for electricity generation and its less negative impact on the environment and the population.

However, fossil fuels have experienced a renaissance in recent years, thanks to modern prospecting techniques, which have made it possible to find new deposits that were unknown until a few decades ago.

Although most of the fossil fuels when burned produce a negative impact on the environment and the population, it is expected that the use of this type of energy sources, particularly the use of natural gas for electricity generation, will continue to be an essential component of the energy mix of several countries all over the world. On the other hand, the use of oil and coal for electricity generation it is expected to decrease significantly by 2050, but they will continue to be an essential component of the energy mix in several countries at least until that year.

Keywords: fossil fuels, crude oil, natural gas, coal, electricity generation

INTRODUCTION

The world is in constant evolution. The economic development that sparked the industrial revolution is causing society to evolve in a way never seen before. The computerization of society is making the economic and social development of many countries advance in an accelerated way, facilitating the appearance of new technologies in all industrial sectors.

A fully industrialized nation is a society where its social and economic growth is linked to the use of a different type of energy sources in almost all its economic sectors. In this type of societies, its economic and social development is intimately linked to their energy capacity installed. Therefore, energy sources and their transformation methods constitute an essential element in the degree of economic and social growth for this group of countries.

Humans need energy for whatever function they develop. Houses must be heated, and power is necessary for the development of industry and agriculture, among other human needs and activities. All the processes that provide humans with the luxuries and comforts in their daily life require the use of different energy sources. These sources can be renewable and non-renewable. Until today and for the coming decades, the world moves mostly with non-renewable energies that harm the environment and the population. The most common non-renewable energy source is the so-called "fossil fuels." This type of energy source is threatened with exhaustion due to its high quantity of consumption at world level. If no new deposits of fossil fuels are founded, it may disappear in the coming decades.

According to Collins English Dictionary, fossil fuel can be defined as an organic substance as coal[1], petroleum (oil)[2], and natural gas[3] found underground in deposits formed in a previous geologic period from de remains of dead plants and animals and used as a source of energy in different economic sectors in many countries all over the world. Fossil fuels are available in large quantities in most of the regions. However, the excessive use of fossil fuels in several economic sectors in almost all countries are consuming at high speed all existing resources and that their replacement could take centuries.

Fossil fuels are characterized by releasing in the combustion process, not only the solar energy accumulated over hundreds of millions of years, but also release into the atmosphere vast amounts of CO_2, a gas that intervenes in global warming due to the greenhouse effect. During the combustion process, specific amounts of sulfur dioxide are also released into the atmosphere, which gives rise to the formation of the so-called "acid rains" that affect the biosphere, due to the damage they cause in the forests, rivers, lakes, and arable land.

---

[1] It is mainly carbon found in large deposits in the soil. To extract it, mines are constructed where the coal resources have been found.

[2] It is a mixture of a variety of hydrocarbons in the liquid phase. It is composed of other large impurities and is used to obtain various fuels and by-products.

[3] It is composed mainly of methane gas and small percentage of hydrocarbon as ethane, propane, butane, nitrogen, among others. Natural gas is the less polluting and purer of all types of fossil fuels.

Without a doubt, fossil fuels were in the last decades and will continue to be in the coming years, the primary source of energy that many countries have throughout the world, despite the increased role that renewable energy sources are now playing and will continue to play in the energy mix of many countries during the coming decades[4]. However, according to World Energy Outlook 2018, "major transformations are underway for the global energy sector, from growing electrification to the expansion of renewables, upheavals in oil production and globalization of natural gas markets. Across all regions and fuels, policy choices made by governments will determine the shape of the energy system of the future."

It is essential to recognize that all energy options must be kept open, and no technology should be idolized or demonized when governments and the energy industry discuss the structure of the energy mix for the country.

Fossil fuels contain high percentages of carbon and include petroleum, coal, and natural gas. Fossil fuels range from volatile materials with low carbon to hydrogen ratios like methane[5] to liquids like petroleum, and nonvolatile elements composed of almost pure carbon, like anthracite coal (Wikipedia, 2019). Fossil fuels get their energy directly from the sun. The plants transformed into these fossil fuels used photosynthesis to capture solar energy.

Fossil fuels are fuels that when burned produce different forms of energy, but also some contaminating gases such as CO_2 with a negative impact on the environment and the population[6]. Without a doubt, this type of energy source has been the axis of global economic development since the industrial revolution in the 18th century until today[7]. The reason is

[4] Currently statistics show that at world level fossil fuels provide between 85% and 90% of the energy that the world consume.

[5] Methane can be found in hydrocarbon fields either alone, associated with oil, or in the form of methane clathrates.

[6] It is important to stress that the CO_2 concentration since the beginning of the 20th century has caused an excess of temperature or global warming that has therefore the known climate change, which is affecting several countries all over the world.

[7] Fossil fuels began to be exploited and used since the industrial revolution around the year 1775 when James Watt invented the steam engine. Since then, fossil fuels began to be exploited and used in different economic sectors in several countries all over the world. However, the use of fossil fuels is the main responsible of the high environmental pollution and the negative effects on the population.

straightforward: the calorific value of fossil fuels is exceptionally high, so when burned generate a high amount of heat that can, in turn, transform it into different forms of energy. Fossil fuels are commonly used for electricity generation, heating, water desalination, transportation, rocket fuel, production of lubricants, hydraulic fluid, plastics, and polyester materials, among others.

Fossil fuels were considered in the past as a reliable source of energy and are expected to play an essential role in everyday life at least until 2050. By 2040, the OPEC[8] foresees a 50% increase in energy demand and that fossil fuels are expected to supply more than 75% of the energy mix of countries all over the world. The OPEC does not foresee any scenario in which non-fossil fuels could come even close to overtaking fossil fuels in the decades ahead. For the foreseeable future, the use of renewables for electricity generation will only ever be supplementary, at best, to the use of fossil fuels[9] for this specific purpose. But it is essential to be aware that fossil fuel reserves are limited, and currently, it is consumed at a much higher rate than what is produced. The consequences of this situation, for example, is that if the world continues at the current rate of oil consumption, it is estimated that in about 100 years all the reserves of this type of fossil fuel will have been exhausted (Quiroga Martinez, 2017) unless new oil deposits are found.

Despite the multiple uses of fossil fuels in different economic sectors, many governments and organizations around the world seek to change the current pattern of CO_2 emissions and are adopting several measures to achieve the goal set in the Paris Agreements to keep the global temperature rise below 2°C. To accomplish the objective mentioned above, a significant advance in innovation and development of renewable energies, an increased in energy efficiency, the development of more effectiveness accumulation systems and an increased used of smart grids should be registered in the coming decades. It is expected that before 2050, radical policies aiming at a clean future should be consolidated by supporting the transition to electric

---

[8] OPEC: Organization for the Petroleum Exporting Countries.

[9] Within the energy mix of many countries, natural gas is the cleaner fossil fuel of the future.

vehicles[10], increasing taxes on the use of fossil fuels for electricity generation and heating, specifically coal, and prohibiting the operation of old and ineffective coal-fired power plants in the next decade[11].

As a result of the implementation of the measures mentioned above, energy consumption has slowed down in the Organization for Economic Cooperation and Development (OECD) area[12], and this situation is likely that is not going to change in the coming decades. In the case of coal, the capture and storage of carbon, especially for natural gas-fired power plants, should continue to be part of the energy landscape in many countries due to its flexible availability for electricity generation.

It is important to single out that fossil fuels have experienced a renaissance in recent years, thanks to modern prospecting techniques, which have made it possible to find deposits that were unknown until a few decades ago. In countries such the USA and Canada, among others, the industry places high hopes in the new findings. According to several experts' opinions, the latest trend seems unstoppable, despite the critics that launch the ecologists against the method of prospecting called "fracking." The USA could soon reach the first place as a producer of fuels of fossil origin, even surpassing Saudi Arabia, due to the extending use of this new method of prospecting.

Based on the results of studies carried out by the International Energy Agency, the following can be stated: the share of fossil fuels in the world's primary energy consumption has not changed in the last 30 years, staying between 80% and 85%. This data would reveal that the technological revolution of clean energy and energy efficiency has not been as radical as many experts think. However, to meet the growing world demand for

---

[10] The UK, France, and China have the intention to prohibit the use of gasoline and diesel cars and truck in the coming years. In certain region of Spain, the use of gasoline and diesel cars will be prohibited before 2050, according to the proposals adopted by some regional authorities.

[11] Since the electricity sector is currently the largest source of carbon emissions to the atmosphere, it is essential that the world continue to look for ways to reduce emissions in this sector. In comparison, reductions in carbon emissions from the transportation sector in all scenarios up to 2040 are relatively small.

[12] Energy consumption has shown a significant growth rates in emerging countries, particularly in China and India.

energy, expected to be around 50%[13] if developments follow the trends seen in the past, it is indispensable to increase the efficiency in the use of fossil fuel for electricity generation, heating, and other purposes, and, at the same time, to reduce greenhouse emissions.

According to the BP Statistical Review of World Energy 2018 report, world primary energy consumption reached 13.511 million tons of oil equivalent in 2017. Between 2007 and 2017, world primary energy consumption grew at an average annual rate of 1,5%. World oil consumption (including biofuels) was 4.622 million tons of oil equivalent in 2017, accounting for 34% of the world primary energy consumption. World natural gas consumption was 3.156 million tons of oil equivalent in 2017, accounting for 23% of the world primary energy consumption. World coal consumption was 3.731 million tons of oil equivalent in 2017, accounting for 28% of the world primary energy consumption. It is important to stress that currently, crude oil is the most abundant energy source, accounting for around 39% of all fossil energy sources, followed by coal with 33%, and natural gas with 28% (Ritchie H. and Riser, M., 2018).

Current studies foresee that between 2020 and 2050, half of the world fossil fuel reserves could be exhausted. It is also expected that a high level of annual fossil fuels production will be reached between the years 2020 and 2021. However, the increased use of renewable energy sources for electricity generation and heating could delay the extinction of half of the fossil fuels world reserves.

Despite the negative impact on the environment and the population using fossil fuels for electricity generation and heating, no alternative allows many countries to eliminate the use of crude oil, coal, and natural gas, and to substitute them by renewable energy sources for the mentioned above

---

[13] There is large uncertainty about the level of the future world energy demand. According to Copenhagen Economics (2017), it is expected that future world energy demand could increase anywhere between 30% and 60%. On the other hand, OPEC recognizes as the most likely future scenario a fall in global crude oil demand after 2040. In any case, and according to OPEC source, it is likely that the world crude oil consumption will increase very modest during the coming years from the current 94 million barrels per day to a maximum of 100,9 million barrels within just a decade, to begin then its slow decline.

purpose soon[14]. Europe committed to ambitious decarbonization of its economy and to cut harmful emissions by closing old and the most contaminating coal-fired power plants. The energy industry, cities, and households will need to change the way they operate and are powered. The EU is already a leader in the efficient use of natural resources and the global fight against climate change. European firms hold 40% of the world's patents for renewable energy technologies (EU COM 2017/2025, 2017).

Disregards of the fast growth of renewables in the last decades, fossil fuels are expected to remain as the dominant primary energy source for many countries all over the world at least for the next 20 years. According to the BP Energy Perspectives report, natural gas will become the fastest growing fossil fuel, with the US betting shale gas and liquefied natural gas (LNG). Renewable energies will account for 25% of global primary energy growth in 2035 and more than 33% of global energy generation growth. The costs of renewables are expected to continue to fall during the coming decades. It is likely that, by 2035, the growth of global coal consumption will decline sharply, and some countries will exclude this type of energy source for electricity generation and heating even before that year.

Fossil fuels are today the main source of primary energy and account for 86% of world energy consumption. During the next two decades, it is expected a gradual transition in the energy mix of many countries to cleaner sources of energy. For example, renewables will grow fast to rise from around 3% today to 10% by 2035, whereas natural gas is expected to replace coal as the second-largest fuel source by that year. In 2035, fossil fuels will remain as the dominant sources of energy, accounting for more than 75% of total energy supplies down from 86% in 2016; this means a reduction of 11% (Figuera, 2017). This situation is not expected to change at least until 2050 drastically, but the role of fossil fuels in the energy mix of several countries will be less relevant that it is today.

---

[14] It is important to single out the following: More than a century took oil to replace coal as the world leading energy source. The bulk of the available studies carried out by several experts estimate that crude oil and natural gas will supply approximately 50% of primary energy demand by 2030. In addition, the growth potential of unconventional hydrocarbons, shale gas or shale oil reinforces the hegemony of these sources at least during the coming decades.

Summing up the following can be stated: the choice of the optimal energy mix by any country should be made to guarantee competitiveness, the security of supply, respect for the environment and the population, for the benefit of sustainable development, and the fight against climate change.

Without a doubt, there is a closed relationship between human development and energy consumption. For this reason, any substantial increase in welfare and human development increase the use of energy in any of its different forms. Today, around 80% of the world population lives in countries where the average energy consumption is relatively low and does not cover the entire community. To reduce the percentage mentioned above to 33% by the year 2040, the world would require consuming around 65% more energy than today, and an essential increase in the resources allocated to the energy sector in several countries, particularly developing countries.

THE WORLD CRUDE OIL SECTOR

Crude oil is formed by a mixture of hydrocarbons[15], oxygen, nitrogen, and a small amount of sulfur. It is a liquid fuel that includes all fossil fuels of liquid hydrocarbons and can refer to crude oil or derivative products made from refined crude oil. "Crude oils contain many different types of impurities including toxic, corrosive and reactive substances that can be costly to deal with. One of the most important impurities is sulfur, and crude oils are generally classified according to sulfur content as sweet or sour, sweet crudes typically containing less than 0.5% sulfur and sour crudes more than 2% sulfur. Higher sulfur crude oil requires costly treatment, and hence

---

[15] The world liquid hydrocarbons industry classifies crude oil according to its API density (international parameter of the American Petroleum Institute). The API density is used to classify the different types of crude that can be found in the Earth. The best quality crude oils are those that are classified as "light" and/or "soft" and "sweet." Other types of crude are classified as "intermediate," "heavy" and "extra heavy." According to different sources, most of the world's oil resources correspond to viscous and heavy hydrocarbons, which are difficult and expensive to produce and refine. In general, the heavier or denser the crude oil, the lower its economic value. The lighter and less dense crude fractions, derived from the simple distillation process, are the most valuable.

sour crude oils are generally sold at a significant discount as compared to sweet crude oils" (Clews, 2016).

Crude oil is used for a wide range of applications, including for electricity generation and heating, in the transport sector, to produce plastic and synthetics and other products, among others. Bitumen, the thickest form of petroleum, is used for paving roads, forming the blacktop, and roofing[16]. It is challenging to find a country that does not use crude oil for some of the activities mentioned above.

"World oil production has increased more than 2,5-fold over the last 50 years. The 1970s "oil crisis" resulted in a sudden drop in consumption between 1973 and 1974 following an oil embargo of the Organization of Petroleum Exporting Countries (OPEC) in 1973." Another oil shock following the Iranian Revolution in 1979 and "an 'oil glut' in the early 1980s, where there was a surplus of crude oil as a result of reduced consumer demand following the energy crisis of the 1970s" (Roser H., and Ritchie, M., 2018).

The evolution of crude oil production at world level during the period 2010-2017 is shown in Figure 1.

According to Figure 1, the crude oil production at world level increased by 11,2%, rising from 83.325 thousand barrels per day in 2010 to 92.649 thousand barrels per day in 2017. It is significant to single out that the amount of crude oil that a particular region produces is not always in correspondence to the size of its proven crude oil reserves. For example, the Middle East contained most of the world's crude oil proven reserves but accounted for only about 30% of global crude oil production. The US, by contrast, laid claim to only about 1,5% of the world's proven crude oil reserves but produced about 10% of the world's crude oil (Hudson Teslik, L., and Sparks, K., 2009).

---

[16] Crude oil can be used when it is refined to transform it into a range of useful products for industry and consumers. Why? Crude oil is a highly variable mixture of heavy and light hydrocarbons that need to be separated in a refinery to turn them into usable products. More than 3.000 petrochemical products can be obtained out of this process. With crude oil the following products can also be obtained: gasoline, naphtha, fuel oils, lubricating oils, kerosene, asphalt, and paraffin wax.

It is expected that crude oil production at world level will continue to grow at least during the coming years and to play a huge role in several economic sectors in many countries all over the world.

On the other hand, crude oil consumption at world level increased during the whole period 2010-2017, and it is expected that this trend will continue without change during the coming years. According to BP source, although crude oil demand will grow more slowly in the coming decades, the peak in crude oil demand is expected not to occur until beyond 2040, thanks to a moderate increase in crude oil consumption in emerging countries. Other experts considering that the beginning in the decline of crude oil demand will begin in the 2020s.

It is important to stress that the use of crude oil for electricity generation and heating is today lower than in the past. However, the total consumption of crude oil increased in the past seven years as can be seen in Figure 2.

The evolution of crude oil consumption at world level during the period 2010-2017 is shown in Figure 2.

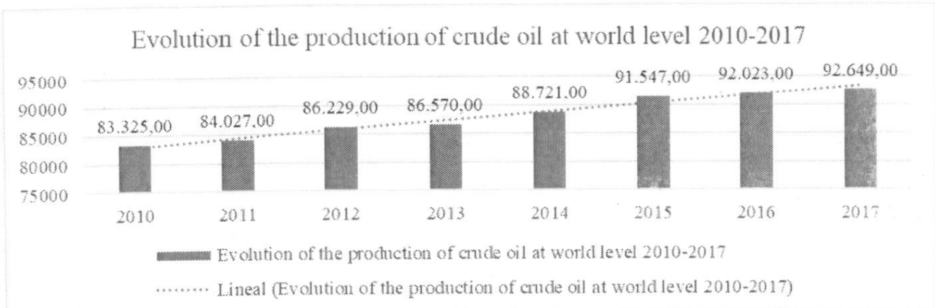

Source: BP Statistical of World Energy 2018 (June 2018).

Figure 1. Evolution of the production of crude oil at world level during the period 2010-2017.

According to the data included in Figure 2, the following can be stated: the consumption of crude oil at world level increased by 10,9% during the period considered, rising from 88.535 thousand barrels per day in 2010 to 98.186 thousand barrels per day in 2017. After looking at Figure 2, it is easier to state the following: crude oil consumption at world level increased each one of the years included in the period considered. Based on this fact, it can be expected that the current trend in the use of crude oil will continue to increase at least during the coming years, mainly in the transportation sector. The consumption of crude oil for electricity generation and heating in many countries will continue to decrease during the coming decades.

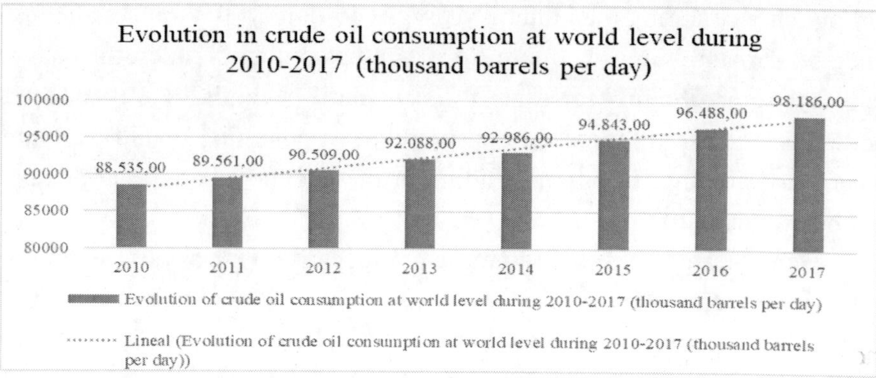

Source: BP Statistical Review of World Energy 2018 (June 2018).

Figure 2. Evolution of crude oil consumption at world level during the period 2010-2017.

The single largest world crude oil consumer is the US, followed by China, and India. By region, the largest crude oil consumers are Asia and Pacific, followed by North America, and Europe and Eurasia (BP Statistical Review of World Energy, 2017).

Several factors could reduce the crude oil demand growth rate in the coming years in many countries, particularly in developed countries. Among these factors, the following are considered the most important ones:

- Decline in population in several countries, particularly within the OECD countries;
- Adoption of new energy efficiency and profitability measures by governments and the industry sector in several states that use crude oil in their productions;
- Increase dependency on the use of renewable energy sources for electricity generation and heating;

- The advance of energy saving in all economic sectors in many countries;
- The significant commitment to electric cars and the increased use of biofuels for electricity generation and heating;
- The increased use of natural gas in several economic sectors in many countries all over the world;
- New advances in batteries that allow great electricity storage.

It is important to note that approximately 63% of today world crude oil consumption comes from the transport sector, and this percentage is expected to increase at least during the coming years. As a result of the different measures adopted by several countries to reduce the negative impact that the use of oil for electricity generation has on the environment and the population, the use of crude oil for electricity generation at the world level is expected to continue to fall until 2050 and, in several of them, beyond that year.

According to the World Energy Resources 2016 report, "oil remained the world's leading fuel, accounting for 32,9% of global energy consumption." However, it is likely that crude oil participation in the world energy mix will be lower in the future than today, particularly for electricity generation and heating, but will continue to be a vital energy source in the transportation sector. By region, crude oil demand in OECD North America, in 2016, was the highest among all regions with 24,3% of world crude oil demand, followed very closely by Asia with 23,8%. It is predictable that this situation will change before 2020. In that year, it is expected that Asia will have the highest crude oil demand at world level with 25,2% of the world total, an increase of 1,4% concerning 2016, followed by the North America region with 24,5%, an increase of only 0,2% with respect the level registered in that year. Emerging economies now account for 58,1% of world energy consumption and global demand for liquid hydrocarbons is expected to continue to grow during the coming years (BP Statistical Review of World Energy 2016, 2016).

The evolution of the world proven crude oil reserves by regions during the period 2010-2017 is shown in Table 1.

Table 1. Evolution of the world proven crude oil reserves by regions during the period 2007-2017 (million barrels)

Country/Region	2007	2016	2017	R/P ratio
North America	221.500	227.700	226.100	30,8
Latin America	125.300	328.900	330.100	125,9
Eastern Europe and Eurasia	145.300	144.900	144.900	27,8
Western Europe	15.100	13.400	13,400	10,4
Middle East	754.900	807.700	807.700	70
Africa	119.700	126.500	126.500	42,9
Asia and the Pacific	45.300	48.300	48.000	16,7
Total	1.427.100	1.697.100	1.696.000	50,2

Source: BP Statistical Review of World Energy 2018 (June 2018).

According to Table 1, the world proved crude oil reserves increased 18,8% during the period 2007-2017 rising from 1.428.100 million barrels in 2007 to 1.696.000 million barrels in 2017. In that year, the substantial crude oil reserves (47,6% of the world total) were in the Middle East, followed by the Latin American region with 19,5% of the world total. Venezuela is the country that has the world most significant crude oil reserves estimated at more than 300 billion barrels. It is important to stress that, in 2017, the crude oil reserves in the Middle East represents 65,36% of the OPEC total crude oil reserves[17]. At the same time in that year, the crude oil reserves in OPEC member states represent 81,89% of the total world crude oil reserves. If the crude oil reserves and production level registered in 2015 is taken as a reference, then it can be confirmed that the number of years of oil reserves left reach 50,7 years (Ritchie H. and Roser, M., 2018). In 2017, the number

[17] Since 2007, the Middle East region (including Iran but not North Africa) laid claim to about 56% of the world's total proven crude oil reserves—that is, more than the rest of the world combined. Following the Middle East were Canada and the US, Latin America and the Caribbean, Africa, and Russia. These regions contained less than 15% of the world's proven crude oil reserves (Hudson Teslik L. and Sparks H., 2009).

of years of world crude oil reserves reached 50,2 years (BP Statistical Review of World Energy, 2018.

The world crude oil demand by regions during the period 2014-2018 is shown in Table 2.

From Table 2, the following can be stated: the growth in crude oil demand within the North America region during the period under consideration was 1,6%, rising from 24,1 million barrels per day in 2014 to 24,5 million barrels per day in 2018. The other significant growth in crude oil demand was registered by "other Asia" with 13,2%, followed by China with 10,5%. OECD Europe and OECD Asia-Ocean are the only two regions reporting a decrease in crude oil demand during the period considered.

Table 2. World crude oil demand by regions during the period 2014-2018 (million barrels per day)

Region	2014	2015	2016	2017	2018
OECD Americas	24,1	24,2	24,3	24,4	24,5
OECD Asia- Ocean	8,1	8,0	7,9	7,9	7,9
OECD Europe	13,4	13,3	13,3	13,2	13,1
Another Europe	0,7	0,7	0,7	0,7	0,7
China	10,4	10,6	10,9	11,2	11,5
Other Asia	12,1	12,5	12,9	13,3	13,7
Latin America	6,8	6,9	7,0	7,1	7,2
Middle East	8,1	8,3	8,5	8,8	9,0
Africa	3,9	4,1	4,2	4,4	4,5
Others	4,8	4,6	4,7	4,7	4,8
World	92,4	93,3	94,5	95,7	96,9

Source: World Energy Resources 2016, used by permission of the World Energy Council, www.world energy.org.

According to the US Energy Information Administration (EIA) database, the world economy consumes more than 90 million barrels of crude oil every day. Around 61% of all global crude oil demand, up to a total of 56,5 million barrels, is transported by sea to the purchasing countries. The main crude oil maritime route is the one used to export crude oil from the Persian Gulf countries. Across the Strait of Hormuz, which separates Iran and Oman, 17 million barrels pass each day, the majority of which goes to the Asia-Pacific region (12 million barrels per day or 70,6% of the total). The remaining five million barrels circulate mostly through the Suez Canal to Europe and in a smaller proportion to America around Africa. The routes from Russia and other countries of the former Soviet bloc that go to the Mediterranean and the Baltic, transport about 3 million barrels per day for each of these two routes.

Despite all efforts made until today by the oil industry to reduce the use of crude oil for electricity generation and other purposes, the substitution of crude oil to meet the energy needs of different sectors of the economy in many countries is not going to happen anytime soon. It is not expected to reach more than 5% over the next five years (World Energy Resources 2016, used by permission of the World Energy Council, www.worldenergy.org, 2016).

It is also important to single out that more than half of the increase in global energy consumption is used for power generation due to the measures adopted towards global electrification at world level. For this reason, the share of energy used for power generation will rise from 42% today to 45% by 2035[18]; this represents an increase of 3% for the whole period under consideration. In the specific case of crude oil, the growth in the global consumption of liquid fuels is going to be carried out by the transport sector (63% today to 88% in 2035) and by the industry sector, especially in petrochemicals, with an increase of 1,2 million barrels per day in 2035. The transport sector will account for almost two-thirds of the rise (BP Energy Outlook 2016, 2016). In 2016, crude oil imports at world level reached 2.117,8 million tons (BP Statistical Review of World Energy 2017, 2017).

Summing up the following can be stated: the demand for petroleum products has changed significantly since the mid-20th century. In industrialized countries, the need of light crude oil products, such as fuel, has increased considerably, while the consumption of heavy fuel oil has declined, as well as of other energy sources, for heating and power

[18] It is important to single out that electricity generation and heating is the main sector where all types of energy sources compete. Without a doubt, renewables and natural gas are the two types of energy sources with the major impact in the structure of the energy mix of many countries, surpassing coal in most of them, due to their minimum negative impact on the environment and the population. This situation will not change at least until 2040.

generation[19]. The Chinese slowdown and the US shale industry are going to halt any oil rally, whose prices will remain low for ten more years, according to Vitol Group, the world's largest independent oil trader. According to the opinion of this group, it is hard to see a dramatic crude oil price increase in the coming years. It is expected that crude oil prices will move around US$50 per barrel during the next decade. These forecasts would mean, according to different expert's calculations, that the oil-rich countries and the energy industry will be facing the most extended period of low prices since the period 1986-1999 when crude oil moved basically between US$10 and US$20 dollars a barrel. Today, one of the problems of having a low crude oil price is that there is too much supply[20] at a time when the global economy is much more efficient and, for this reason, consume less crude oil than in the past.

It is projected that crude oil demand will grow in the coming years before gradually stabilizing, while global coal consumption will remain relatively stable. However, in all possible scenarios, continued investment in new crude oil reserves will be necessary to satisfy the projected increase in demand for this type of energy source in 2040.

THE WORLD NATURAL GAS SECTOR

Natural gas is, without a doubt, a real alternative to oil dependence, releases 40% less CO_2 in its combustion than coal and 20% less than petroleum by-products, and does not emit sulfur or solid particles. Natural gas is the non-renewable energy more compassionate with the environment and the population. The main characteristics of natural gas are the following:

---

[19] "Lighter crudes are normally more valuable than heavier crudes as a greater volume of more useful lighter components can be produced with less processing cost" (Clews, 2016).

[20] The production of shale oil by the US is another reason of having today a low crude oil price. The US, typically a massive importer of crude oil and natural gas in the past, it is expected to be independent from foreign sources in 2020, thanks to the prospecting of fracking.

- Its extraction is in deposits;
- It is composed of methane, carbon dioxide, nitrogen, hydrogen sulfide or helium, ethane, propane, butane, and other alkanes;
- It is a very versatile fuel and with less CO_2 emissions in its combustion in comparison to oil and coal[21];
- Its uses are as domestic gas, in vehicular natural gas, heating, in power plants and cogeneration, fuel cell, among many others;
- It produces fewer greenhouse gases.

Among the main applications of natural gas are the following:

- Metal melting;
- Electricity and heating generation;
- Production of petrochemicals;
- Steam generation.

Natural gas[22] tops the list of fossil fuels when it comes to burning cleanly. As the process of combustion of natural gas is almost always complete, the level of bioproducts that are disseminated in the environment is very low. Besides, technological advances have massively reduced nitrogen oxide, a common contaminant of natural gas when it is used in different economic sectors.

Natural gas is the second most-used energy source for electricity generation and heating at a world level, and it is expected to be the only fossil fuel whose participation in primary energy consumption is likely to grow during the coming years. The reason is straightforward; natural gas is the only fossil fuel with the lowest level of pollution it produces in the

[21] Natural gas emits only around 50% of the CO_2 generated by the combustion of coal and around 75% of the CO_2 emitted by the combustion of crude oil. It is for this reason that natural gas began to be used in both the domestic and industrial sectors of many countries, and its use has led to profound changes in the economy of these countries. This change is due to the calorific value of the natural gas (88.8 MJ/m^3), its cleanliness, its comfortable use, and its easy and economical distribution through long gas pipelines that amortize in a few months of exploitation the initial investment.

[22] If the temperature is very high (more than 150° C), the oil can turn into natural gas.

environment and the population in comparison with oil and coal. For this specific characteristic of natural gas, the use of this type of fossil fuel is expected to play an important role in the global energy transition towards a cleaner, cheaper, and safer energy during the coming decades.

Despite what has been said above, the future of the world natural gas demand is highly uncertain. To overcome this uncertainty, the adoption of new energy policy and relevant measures to reduce costs is indispensable to make natural gas more competitive concerning other energy sources used for electricity generation and other purposes, such as renewable. The construction of the necessary infrastructure to ensure the use of natural gas in the most economical manner, the support of governments, and the closing of regulatory gaps, among others, are necessary actions that should be carried by governments and the energy industry to unlock the socioeconomic and environmental benefits related to the production and consumption of natural gas.

Relevant data on natural gas reserves and production at world level and by region is included in Table 3.

Table 3. Natural gas reserves and production at a world level and by region in 2017

Region	Proved reserves (trillion cubic feet) 2017	Production (trillion cubic feet) 2017	R/P ration (years)
Africa	487,8	5.006,96	61,4
Asia Pacific	681,8	27.174,58	31,8
Europe and Eurasia	2.195,5	39.063,45	84,8
Latin America and the Caribbean	290,3	6.122,75	45,9
Middle East	2.794,2	18.943,82	119,9
North America	381,9	33.290,27	11,4
Total	6,831,7	129.601,82	56,2

Source: BP Statistical Review of World Energy 2018 (June 2018).

According to Table 3, the North America region has the lowest R/P ration (11,4 years) in comparison to other areas included in the above table. This weak relationship impacts negatively on the level of production of natural gas in the region. The region with the maximum R/P ration is the

Middle East with 119,9 years, followed by Europe and Eurasia with 84,8 years, and Africa with 61,4 years. The Middle East is the region with the most significant natural gas reserves with 2.794,2 trillion cubic feet, followed by Europe and Eurasia with 2.195,5 trillion cubic feet, and the Asia Pacific with 681,8 trillion cubic feet. Europe and Eurasia is the region with the highest natural gas production with 39.063,45 trillion cubic feet, followed by North America with 33.290,27 trillion cubic feet, and the Asia Pacific with 27.174,58 trillion cubic feet.

The world consumption of natural gas is expected to increase up to 2040 rising from 120 trillion cubic feet in 2012 to 203 trillion cubic feet in 2040; this means an increase of 69,2%. By type of energy source, natural gas is expected to register the most significant increase in the world's primary energy consumption at least until 2040 (International Energy Outlook 2016, 2016).

In several countries, natural gas continues to be an essential fossil fuel used for electricity generation and in the industrial sector. For electricity generation, natural gas is for many countries an attractive option, and, for this reason, these countries are supporting the construction of new natural gas-fired power plants. It is important to stress that natural gas-fired power plants need moderate capital investment to be constructed and, for this reason, the production of this type of power plants can be sold to an attractive price. Also, natural gas-fired plants operate with a relatively high level of efficiency in comparison with other types of power plants, and natural gas is the less contaminated fossil fuel that can be used for electricity generation and other purposes. For the reasons mentioned above, natural gas can be used, without difficulty, instead of coal and other liquid fuels for this specific purpose (International Energy Outlook 2016 with Projection to 2040, 2016).

Based on what has been said in the above paragraph, the following can be stated: world natural gas producers are expected to increase the supply of this type of energy source by 69,2% to meet the growing demand for natural gas projected until 2040. It is also likely that the most substantial increases in natural gas production will occur in Asia, especially in those countries outside the OECD (18,7 trillion cubic feet), followed by the countries in the Middle East (16,6 trillion cubic feet), and in the OECD America (15,5 trillion cubic feet) (International Energy Outlook 2016 with Projection to 2040, 2016).

The evolution in the production of natural gas at world level during the period 2010-2017 is shown in Figure 3.

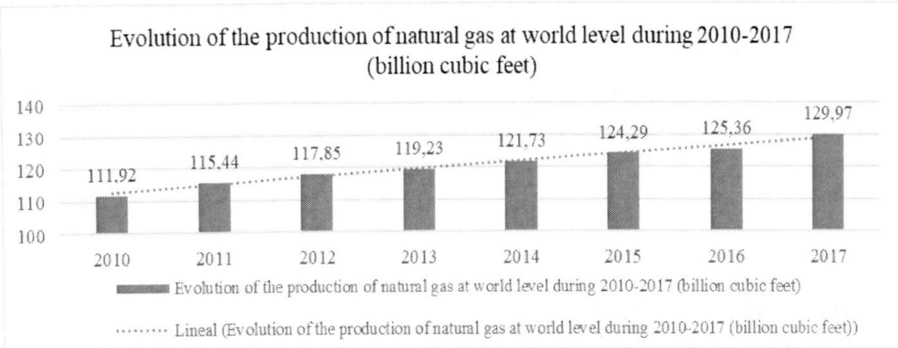

Source: BP Statistical Review of World Energy 2018 (June 2018).

Figure 3. Evolution of the production of natural gas at world level during the period 2010- 2017.

According to Figure 3, the production of natural gas at world level during the period 2010-2017 increased by 16,1%, rising from 111,92 billion cubic feet in 2010 to 129,97 billion cubic feet in 2017. It is relevant to single out that in each of the years within the period considered the production of natural gas increased. The most significant increased was registered between 2016 and 2017 (3,7%). Based on the data included in Figure 3, it is expected that the production of natural gas at world level will continue to grow during the coming decades.

On the other hand, and taking into account the replacement of oil and coal for natural gas for electricity generation and heating in many countries all over the world, the consumption of natural gas at world level increased during the last seven years.

The evolution in the consumption of natural gas at world level during the period 2010-2017 is shown in Figure 4.

Based on the information included in Figure 4, the following can be stated: The consumption of natural gas increased by 15,6% during the period 2010-2017, rising from 112,16 billion cubic feet in 2010 to 129,62 billion cubic feet in 2017. It is likely that this trend will continue without change during the coming years, as a result of the measures adopted by several countries all over the world to increase the use of natural gas for electricity generation and heating, and to reduce the use of crude oil and coal for this specific purpose.

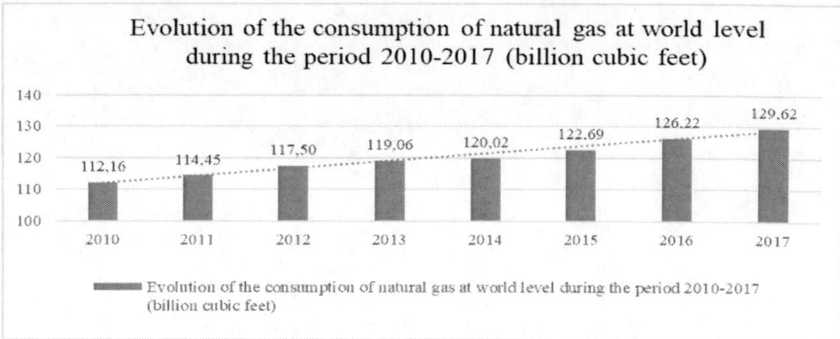

Source: BP Statistical Review of World Energy 2018 (June 2018).

Figure 4. Evolution in the consumption of natural gas at world level during the period 2010-2017.

According to the statistics of the International Energy Agency (IEA), up to 66,7% of the world demand for natural gas is consumed within the producing countries themselves, and the remaining 33,3% is destined for exports. Gas pipelines distribute the vast majority trade of natural gas between countries (70%), and only 30% corresponds to the maritime transport of LNG. In 2017, the total pipeline exports of LNG at world level reached 26.154,12 trillion cubic feet. The countries with the most significant pipelines exports of LNG was the Russian Federation with 7.605,77 billion cubic feet, followed by Norway with 3.855,85 billion cubic feet, and Canada with 2.849,52 billion cubic feet. The total LNG exports at world level reached, in the same year, a total of 13.890,95 billion cubic feet. The countries with significant LNG exports are the following: Qatar with 3.651,05 billion cubic feet, followed by Australia with 2.680,03 billion cubic feet. The significant LNG importers by pipelines at world level are Germany with 3.347,39 billion cubic feet.

It is expected that by 2020, around 40% of the world natural gas trade will be distributed by ship in the form of LNG. On the other hand, the world LNG trade, it is projected to grow 242% during the period 2012-2040, rising from 12 trillion cubic feet in 2012 to 29 trillion cubic feet in 2040. Most of the expected increase in liquefaction capacity is probable to occur in Australia and North America, where many new liquefaction projects are included in the energy sector development plans or are currently under construction. Many of these projects are expected to come online in the next decade. Despite the projection of a steady growth in LNG trade, natural gas flow through oil pipelines is expected to continue to account for most of the world natural gas trade (International Energy Outlook 2016 with Projection to 2040, 2016), for which the construction of new long-distance pipelines and expansion of existing infrastructure until 2040 will be needed. The most significant volumes of natural gas pipeline trade at world level are currently between Canada and the US, and within Europe. It is likely that his situation will not change during the coming years.

THE WORLD COAL SECTOR

Coal is the third of the three fossil fuels used for electricity generation[23] and, at the same time, it is the most contaminated of them when it is used for this purpose. Despite being the most contaminating fossil fuel used at world level for electricity generation and heating, coal has met nearly half of the world's energy demand growth over the past decade. Almost all of this growth was attributable to a rising coal demand from China to satisfy its increasing energy demand. It is important to highlight that two-thirds of the world coal consumption is used for electricity generation and heating; the remainder is mainly used for steel manufacturing (Canada's Energy Future 2016. Energy Supply and Demand Projections to 2040, 2016).

[23] Coal is also indispensable to produce iron and steel; almost 70% of steel production comes from iron made in blast furnaces with the help of coking coal.

According to the International Energy Outlook 2016 with Projection to 2040 report, coal is expected to be the world's slowest-growing energy source during the period 2012-2040, rising by an average 0,6% per year, from 153 quadrillions Btu in 2012 to 180 quadrillion Btu in 2040. This foreseeable increase is considerably lower than the 2,2% per year growth average registered over the past three decades, and the second-largest energy source worldwide—behind petroleum and other liquids—until 2030[24]. From 2030 through 2040, coal is expected to be the third-largest fossil fuels, behind both oil and other liquid fuels, and natural gas.

According to Castro (2018), coal is classified according to its calorific value, which is the power to produce combustion or burn. These are the different types of coal:

- Peat;
- Lignite;
- Mineral Coal;
- Anthracite[25].

World proved coal reserves represents about 70% of the world's currently known fossil fuel energy reserves and is the most widely used in the production of electricity worldwide. Now coal reserves are sufficient to meet 153 years of the global output, roughly three times the R/P ratio for oil and natural gas. By region, Asia Pacific holds the most proved coal reserves with 41% of the world total or 424.234 million tons, followed by Europe and Eurasia with 31,3% of the world total or 323.633 million tons, and the North America region with 25% of the world total or 258.709 million tons. The US remains the largest proved coal reserve holder with 24,2% of the world total

[24] "Coal was the first of the fossil fuels to go into widespread use, displacing low-energy firewood as the leading source of fuel in the US, and triggering the country's industrialization in the second half of the 19th century. Within a few decades, the US went from a net importer of coal (mostly from the UK) to a major exporter of this specific fossil fuel, a development made possible by mining the nation's vast reserves of coal" (Oil 2017, 2017).

[25] It is a carbonaceous substance that is obtained from coal, especially from mineral coal, using the dry distillation method. It is obtained by heating the coal in the absence of air in special furnaces. The coke burns without flame and has a great calorific power. It is used as a fuel, and as a reducing agent in the production of iron in high furnaces (metallurgical coke).

or 250.916 million tons, followed by the Russian Federation with 15,5% of the world total or 160.364 million tons, Australia with 14% of the world total or 144.818 million tons, and China with 13,4% of the world total or 138.819 million tons. Eurasia has the highest regional R/P ratio with 397 years, followed by North America with 335 years, and Europe with 159 years (BP Statistical Review of World Energy 2018, 2018).

The evolution of proved coal reserves by region during the years 1997, 2007, and 2017 are shown in Figures 5, 6, and 7.

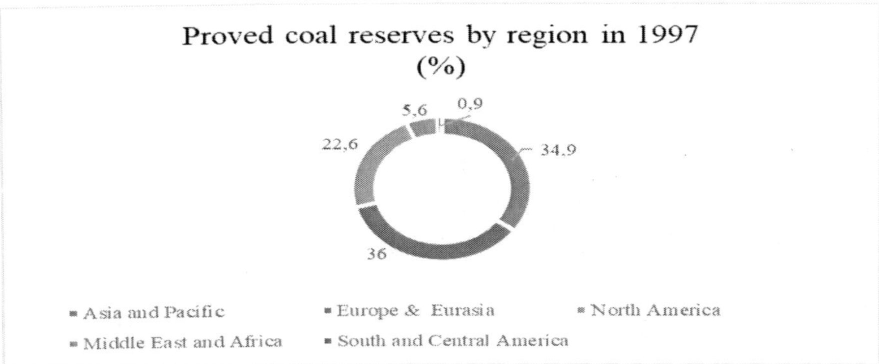

Source: BP Statistical Review of World Energy 2018 (June 2018).

Figure 5. Proved coal reserves during in 1997.

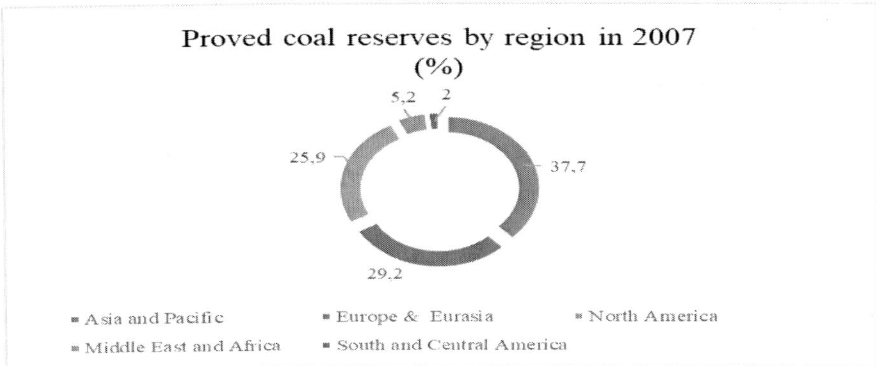

Source: BP Statistical Review of World Energy 2018 (June 2018).

Figure 6. Proved coal reserves in 2007.

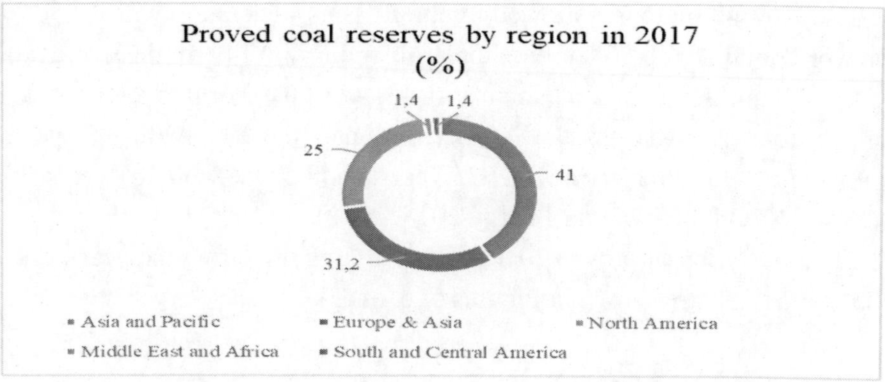

Source: BP Statistical Review of World Energy 2018 (June 2018).

Figure 7. Proved coal reserves in 2017.

According to Figures 5, 6, and 7, the world proved coal reserves decreased by 6,5% during the period 1997-2017, falling from 1.106.210 million tons in 1997 to 1.035.012 million tons in 2017. According to the BP Statistical Review of World Energy 2018 report, the changes in the word proven coal reserves during the period 1997-2017 were the following:

- Asia Pacific: An increase from 34,9% in 1997 to 41% in 2017;
- Europe and Eurasia: A decrease from 36% in 1997 to 31,2% in 2017;
- North America: An increase from 22,6% in 1997 to 25% in 2017;
- The Middle East and Africa: A decrease from 5,6% in 1997 to 1,4% in 2017;
- South and Central America: An increase from 0,9% in 1997 to 1,4% in 2017.

The evolution of world coal production during the period 2010-2017 is shown in Figure 8.

Based on the information included in Figure 8, the following can be stated: coal production increased 4,6% during the period 2010-2017 rising from 3.601,60 million tons of oil equivalent in 2010 to 3.768,60 million tons of oil equivalent in 2017. However, it is predictable that the production of coal at world level will decrease during the coming years, as a result of the implementation of the Paris Agreement on climate change by many countries, and the adoption of specific measures to reduce the use of coal for electricity generation and heating in the US, and China, among others. One of the actions that must be adopted by these two countries, among others, is to reduce the negative impact on the environment and population by the closure of old and inefficient coal-fired power plants used for electricity generation and heating during the coming years[26].

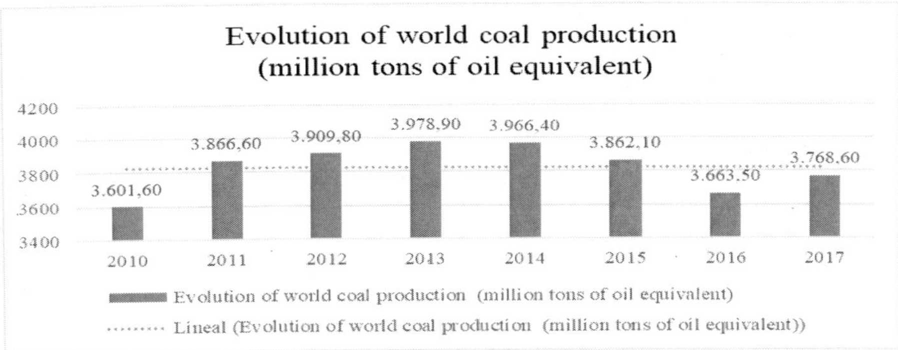

Source: BP Statistical Review of World Energy 2018) (June 2018).

Figure 8. Evolution of the world coal production during the period 2010-2017.

There is a significant regional variation expected to be registered in the outlook for coal production until 2040. The most likely changes are the following:

- Large projected increases for India;
- Considerable growths in coal production in Africa and Russia;
- A growth that slows down to decrease gradually after 2025 in China;
- Few changes in the OECD Europe countries and the US.

---

[26] During the 2010s, the US closed 60 coal-fired power plants. However, the country is still operating around 700 coal-fired power plants and satisfy nearly 33% of the country's energy needs (Beck, 2017).

Asia and the Pacific are, according to BP Statistical Review of World Energy 2018 report, currently the most important markets for coal trade, consuming in 2017, a total of 74,5% (2.780 million tons of oil equivalent) of the world's coal production (3.731,5 million tons of oil equivalent).

During the period 2018-2040, the three largest coal consuming countries are expected to be China, the US, and India, which together it is supposed to consume more than 70% of the world's coal production[27]. In 2017, China accounted for 50,7% of the total coal consumption across the globe (1.892,6 million tons of oil equivalent). As a result of a decrease in the growth of China's economy and the adoption of measures to reduce air pollution and the negative impact on the environment and the population, the use of coal for electricity generation and heating in China was cut during the last years. India is expected to consume 424 million tons of oil equivalent or 11,4% of the world total, followed by the US with 332,1 million tons of oil equivalent or 8,9% of the world total.

The evolution of the consumption of coal at world level during the period 2010-2017 is shown in Figure 9.

According to Figure 9, the consumption of coal at world level increased 3,5% during the period considered, rising from 3.605,6 million tons of oil equivalent in 2010 to 3.731,5 million tons of oil equivalent in 2017. For a better analysis of the behavior in the consumption of coal at world level, the period considered can be divided into two. The first period 2010-2013, the use of coal increased by 7,2% rising from 3,605,6 million tons of oil equivalent in 2010 to 3,865,3 million tons of oil equivalent in 2013. But during the second period 2014-2017, the use of coal decreased by 3,4% falling from 3.862,2 million tons of oil equivalent in 2014 to 3.731,5 million tons of oil equivalent in 2017. The peak in the consumption of coal within the whole period considered was reached in 2013. Several measures adopted by the highest coal consumers countries are responsible for this reduction. The primary purpose of these measures is to reduce, as much as possible, the adverse impact on the environment and the population as a result to the use of this type of energy source for electricity generation and heating.

[27] In 2016, China, US, and India consumed 71,2% of the world coal consumption.

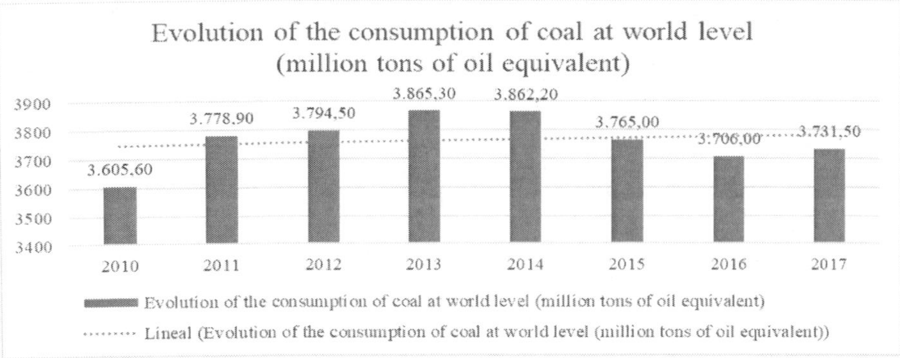

Source: BP Statistical Review of World Energy 2018 (June 2018).

Figure 9. Evolution of the consumption of coal at world level during the period 2010-2017.

Despite the significant reduction in the consumption of coal for electricity generation and heating registered in the last years at world level, coal still is used to produce around 40% of the world electricity production. However, its negative impact on the environment and the population, and its responsibility with the climate change that is taking place worldwide is forcing many countries to speed up the transition to the use of cleaner forms of energy for electricity generation and heating. For this reason, it is likely that the production of coal at world level will continue to decrease during the coming years, mainly due a significant reduction in the consumption of coal by China, the US, and other higher coal consumer countries.

Summing up the following can be stated: global coal consumption is projected to remain roughly the same between 2015 and 2040, with decreasing consumption in China and the US that will offset growth in India. China is expected to continue to be the largest single consumer of coal until 2040, despite a steady decline in the consumption of this type of energy source throughout the period considered. On the other hand, it is also likely that coal consumption in India will continue to grow at an average of 2,6% per year between 2015 and 2040, surpassing the US as the second largest consumer of coal before 2020. In OECD countries, it is projected that the consumption of coal decreases by an average of 0,6% per year during the period 2015-2040, due to the growing competition from natural gas and renewable energies, and a moderate increase in the demand for electricity. Africa, the Middle East, and other non-OECD countries in Asia are predicted to increase the capacity and coal consumption for electricity generation and heating by 2040 gradually.

Finally, it is significant to stress that coal is traded all over the world. In 2016, according to the World Coal Association, overall international trade of this specific type of fossil fuel reached 1.333,5 million tons. Even though this is a substantial amount of coal, it accounts only for about 18% of total consumption because most of the coal produced is used in the country.

ENVIRONMENTAL CONSIDERATIONS

Although fossil fuels allowed a productive development never known in the history of man, there was also a high negative impact on the environment, and the population caused using this type of energy sources for electricity generation, heating, and other purposes. The combustion of fossil fuel generates emissions of different gases such as carbon dioxide, carbon monoxide, among others. These gases have contributed, and continue to provide, an increase in the greenhouse effect, the production of acid rain, and the pollution of air, soil, and water. Polluting effects are not only linked to its combustion of fossil fuels but also to transport them (oil spills) and the byproducts they cause (hydrocarbons and toxic derivatives). The situation is aggravated as a result of the significant growing demand for energy, goods, and services, and due to the increase in the world population and consumption patterns.

Without a doubt, the use of fossil fuels generates pollution[28], but when they burn impurities are released into the air having the following consequences for the environment and the population:

---

[28] It is a reality that much of the pollution from fossil fuels began to occur when the industrial revolution exploded at its peak and has been increased since then.

- Pollute the ecosystem: The burning of fossil fuels produces waste products due to the impurities present in the fuel such as sulfur dioxide, nitrogen dioxide, and volatile organic compounds. Air pollution comes from several sources, but automobiles and industry are the two most significant contributors;
- Weaken the ozone layer. The damaging consequences of the burn of fossil fuels are a reduction of the ozone layer and an increase in the temperature. The decrease in the ozone layer threatens human health, vegetation, and the marine ecosystem. It contributes to the rise of the sea waters, which threaten the coastal regions, particularly small islands;
- Responsible for global warming and climate change: The burning of fossil fuels contributes to the accumulation of greenhouse gases, which is considered by many experts the main factor of climate change and global warming. The most significant contributor to greenhouse gases is the burning of fossil fuels. Almost 80% of carbon dioxide emissions come from the consumption and transformation of fossil fuels into energy;
- Biodiversity loss: There is a severe problem in the natural environment, and that is the rapid disappearance of terrestrial ecosystems and all the biodiversity they harbor. According to Moriana (2018), it is projected that, of the more than 47,000 species evaluated, around 36% are in danger of extinction. Although it must be taken into account, that this disappearance is also occurring at an accelerated pace in the aquatic environment. This loss of biodiversity is due to several factors, among which are the following:

    - The overexploitation of the natural environment;
    - The destruction of the species habitats;
    - The pollution of the ecosystem;
    - The introduction of invasive alien species;
    - The effects of climate change;

- The devastation of the environment and the reduction in the quality of life.

Finally, it is essential to know that the refining and extraction process of fossil fuels is dirty and dangerous, and can pollute the water table and entire bodies of water such as rivers and lakes as well.

Some Advantages and Disadvantages in the Use of Fossil Fuels

Fossil fuels are a technology that is well known in all countries over the world. Although it is often thought that this type of fuel source is a way to meet transportation needs, many of the products that are used every day contain items manufactured using fossil fuel technologies. That includes computers, mobile devices, different types of chemicals, and plastics that are used by millions and millions of people for many different purposes. Even renewable have a base built on fossil fuels.

The use of fossil fuels for different purposes, particularly for electricity generation can provoke, in cases when it is not handled correctly, a severe negative impact on the environment and the population. Then, why all countries use fossil fuels for electricity generation and heating? The answer is straightforward: in most countries, they either do not have substitutes for its use for electricity generation and heating, or they are less expensive than the use of other available energy sources for the same purpose.

In comparison with other energy sources, the use of fossil fuels has some advantages but also disadvantages. In general, according to the paper entitled "The Truth about Oil: Pros and Cons of Fossil Fuels (2018)," these are some of the advantages and disadvantages in the use of fossil fuels in comparison with other types of energy sources:

- High Efficiency: It is important to stress that each ounce of fossil fuels provide more energy than virtually any other types of energy sources;
- Electricity Generation and Heating: Fossil fuels can generate enormous amounts of electricity and heating in comparison with any other type of energy sources;
- Low Cost Per Unit: The cost per watt of power produced using fossil fuels remains quite lower, even when the price of oil gets relatively high;
- Stability in the Production of Energy: Fossil fuels are highly stable in electricity production when compared to other available energy sources;
- Existing Infrastructure: Since fossil fuels have been used for an extended period, there is a vast infrastructure in place to keep things moving. These include hips, pipelines, refineries, and other systems. All these components have been designed and built with crude oil and other fossil fuel products in mind;
- Research and Development: There has been a lot of resources spent on research and development activities associated with the use of fossil fuels in different economic sectors. Companies supporting these research and development activities want to recover the investment made in those activities as soon as possible;
- Employment: There are a lot of professionals, technician, qualifies workers, among others, working in many facilities associated with the use of fossil fuels in different economic sectors in several countries. The closure of these facilities could increase the rate of unemployment in these countries;
- Environmental Implications: More than any form of energy sources, there are serious issues associated with the negative environmental impact caused by the burn of fossil fuels for electricity generation and heating. Also, the extraction process associated with this type of energy source is a dangerous activity;
- Limited Resources: Fossil fuels are not renewable and, for this reason, will eventually run out in the future, depending on the rate

of consumption and the level of new discovering. For this reason, it is complicated to predict when this will occur for each of the fossil fuels available today. However, according to the estimation made by several experts, it is expected that given the rate at which fossil fuels are consumed, and the prediction of discoveries of new deposits, this type of energy source will be exhausted within 100 or 200 years[29]. According to World Energy Council 2016, the regions with the highest coal recoverable reserves are the following: Europe with 30%, followed by North America with 28,2% and East Asia with 13,5%. The regions with the highest crude oil recoverable reserves are the Middle East and North Africa with 48,9%, followed by Latin American and the Caribbean with 21,3%, and North America with 15%. In the case of natural gas, the regions with the highest recoverable reserves are the Middle East and North Africa with 46%, following by Europe with 19%, and South and Central Asia with 12%;

- Carbon Release: The burn of fossil fuels is the primary cause of the increase of CO_2 levels in the atmosphere, in addition to producing other forms of pollution. The contaminating gases produced by the burn of fossil fuels are provoking serious climate changes all over the world, and are affecting the lives of millions of people in almost all countries. Scientists claim that if current levels of CO_2 do not decrease in the coming years, then Earth will become warmer, and this situation will cause severe climate changes affecting many countries all over the world;
- Accessibility to Reserves: The accessibility to reserves is not very complicated yet from the technical point of view. This means that, since extraction is not too complicated, the economic exploitation costs of these reserves are acceptable. However, the main deposits of fossil fuels are concentrated in specific regions on the planet; they are not everywhere. The control and exploitation of these deposits are the cause of conflicts, wars, and social and political tensions.

[29] According to different sources, an estimated of 18 million barrels of oil is consumed in the US every day.

Many of the wars in recent years have been caused by the control of these energy resources, on which many countries depend;

- Transportation and Storage: The transport and storage of some fossil fuels are cheap and easy to handle if compared to other energy sources. Unlike renewable energy, the transport and storage of fossil fuels is easier;
- Production of Fossil Fuels: The human being cannot produce any of the known fossil fuels, and must burn them to use it in the form of electricity or to produce heating, among other purposes;
- Unbalanced Consumption: To the evident differences between developed and developing countries another fact should be considered. This fact is the following: 25% of the world population that lives in the richer countries consumes 75% of the total primary energy produced by the burn of fossil fuels at world level;
- Fossil Fuels are Unsafe: Fossil fuels need to be burned to generate energy, and, for this reason from time to time accidents happen. This type of accidents is usually catastrophic for the environment and the population.

The following are some of the main advantages in the use of fossil fuels to produce energy and other purposes:

Crude Oil

- Crude oil has high calorific power in comparison with other types of energy sources;
- Crude oil is a regular source of energy, particularly in the transport sector;
- A great variety of products can be obtained from the processing of crude oil in specific installations (refineries);
- Crude oil is a reliable source of energy for electricity generation, but its use has been declining at world level in the last decades due to environmental reasons. It is expected that this trend will continue without change in the coming years, and the role played by this type

of energy source within the energy mix of several countries will be lower than today;

Coal

- Coal contains a large amount of energy in comparison with other types of energy sources;
- Coal can be consumed in the same extraction site;
- Coal transportation costs are lower than the transportation costs of other fossil fuels;
- Coal is safer to transport in comparison with other fossil fuels;
- Coal is abundant, burns quickly, and is an accessible source of energy that can be found in several sites in different countries and regions;
- Coal is a reliable source of energy;
- Coal is considered one of the cheapest types of energy source, and one of the most reliable energy sources for electricity generation and heating;

Natural Gas

- Natural gas has a competitive price in comparison with other types of energy source;
- The burned of natural gas is much cleaner than any other fossil fuels;
- Natural gas is an abundant energy source and can be found in several countries and regions all over the world;
- Natural gas is a reliable source of energy for many countries, particularly for electricity generation and heating.

The following are some of the main disadvantages in the use of fossil fuels to produce energy and other purposes:

Crude Oil

- Crude oil is not a renewable energy source. Its reserves will decrease during the coming decades if new deposits are not found. However, the increased use of different renewable energy sources is reducing the consumption of crude oil in many countries, particularly for electricity generation and heating, and, for this reason, its reserves can be exploited more time;
- The burning of crude oil contributes to the greenhouse effect and acid rain and, for this reason, it has a negative impact on the environment and the population;
- Crude oil is a dangerous fossil fuel and can produce an adverse effect on the environment in case of an accident or if not handled correctly;
- During crude oil extraction part of the oil is poured directly into the space surrounding the prospecting and it is very harmful when it comes to extractions at sea;
- During crude oil transportation, due to the diversity of situations, if an accident occurs, then the environmental impact can be severe. Since the crash of a tanker truck to the sinking of an oil tanker, the effects on the environment and the population living in the surrounded areas are often disastrous;
- During crude oil refining, contamination of the environment can occur in the evacuation of the waste produced in the refineries.

Coal

- The extraction of coal is a dangerous activity and has a negative impact on the environment and the population living in the surrounded of the site, particularly to those that work in the coal mines. The extraction of coal can also contaminate the soil, water, and air in the surrounded areas of the coal mine;
- Coal is not a renewable energy source and will be exhausted in the future. However, the increased use of different renewable energy

sources for electricity generation and heating, and the adoption of measures to reduce the negative impact on the environment and the population as a result of the use of coal as an energy source, is reducing the use of this type of energy source for the purpose mentioned earlier in many countries. This situation could extend the extraction of coal reserves for many more years;

- The combustion of coal generates many environmental problems, mainly if it contains a high sulfur index. Without any doubt, sulfur is one of the major contributors to the greenhouse effect and acid rain at world level;
- Coal-fired power plants operation require a large amount of coal to operate most efficiently;
- "Mining, cleaning, transportation and the disposal of coal ash have probably caused more environmental damage to our world than any other human activity to date" (Reid, 2018);

Natural Gas

- The use of natural gas produces some greenhouse gases that have a negative impact on the environment and the population. Nevertheless, in comparison with oil and coal this negative impact is meager;
- Natural gas is not a renewable energy source, and its reserves can be exhausted in the future. However, the increased use of different renewable energy sources is reducing the use of natural gas in several countries, particularly for electricity generation and heating, and this situation could extend the use of natural gas proved reserves for this and other purposes many more years;
- Natural gas, due to its specific characteristics, is difficult to store;
- Natural gas is highly combustible, and its careless handling can lead to explosions[30] with a severe negative impact on the environment and the population.

[30] The main problem with the use of natural gas is that it is odorless, and no leaks can be detected unless an odor has been added. Another problem is that its gaseous form makes it more difficult to handle than a liquid.

In summary, the following can be stated: although it is true that the production of energy from fossil fuels can be more economical than any other form of energy generation, and their transport and storage more accessible, the pollution produced by the use of oil and coal for electricity generation and heating, is the primary cause of the emission of greenhouse gases[31]. For the reasons mentioned above, the use of oil and coal for this particular purpose is entirely discouraged by many governments all over the world.

THE FUTURE OF FOSSIL FUELS

According to the document entitled "The Future of Fossil Fuels: How to Steer Fossil Fuel Use in a Transition to a Low-Carbon Energy System" prepared for the Energy Transitions Commission by Copenhagen Economics in 2017, energy needs at world level expected to grow between 60% and 80% over the next 25 years, accompanying economic development and growth. It is a fact that the world population will continue to grow during the coming decades and, for this reason, the production of energy from fossil fuels, mainly from natural gas, is also expected to increase. This increase aims to satisfy the projected increase in energy demand. It is important to stress that even though the amount of fossil fuels that can be used for this specific purpose is limited, it is challenging for humankind to meet the expected future energy demand without the use of this type of energy source.

In the absence of a profound energy transition from the use of fossil fuels for electricity generation and other purposes to the use of a cleaner energy source, emissions could grow at world level by two-thirds, to almost 60 billion tons CO_2 per year. For this reason, "an energy transition must achieve rapid improvements in both energy and carbon intensity to prevent this

[31] Many of the costs associated with the use of fossil fuels are obvious: the cost of labor, the cost of transportation, the cost of drilling and refining materials, the cost of building energy-generating facilities, among others. But there are other fossil fuel costs that are not included in the energy bill and are paid by the society such as the degradation of air quality, health problems, damage to the land, environmental issues, water pollution, acid rain, among others.

growth, and even more to reduce emissions levels" (Copenhagen Economics, 2017).

There are several issues under consideration in many countries to reduce the use of the most contaminating fossil fuel for electricity generation and heating and, at the same time, to increase the use of other cleaner energy sources. Two of these issues are the reduction of CO_2 emissions and its negative impact on the environment and the population, and the conservation of fossil fuels limited resources. This can be achieved by the substitution of petroleum-based fuels for biofuels, and the increased use of different renewable energy sources for electricity generation and heating; in some countries by the use of nuclear energy for this specific purpose and for desalination purpose as well.

The following are at least three reasons why many countries cannot stop using fossil fuels for the generation of electricity and heating:

- They cannot cover current and future expected energy demand without the use of all or at least some of the fossil fuels available in the world;
- Environmental policies are not among some government top priorities;
- Their economy cannot afford the transit from the use of fossil fuels to the use of renewable energy sources for electricity generation and heating in a short period.

Several governments are adopting measures to reduce their support for the use of fossil fuels for electricity generation and heating[32]. Others, under the pressure of the energy industry, are supporting the use of fossil fuels in different ways, for example, by support specific exploration and extraction projects, organizing official visits to individual countries to secure the raw

[32] Public financial support for the use of fossil fuels reaches one trillion euros at world level in the last years. According to the paper entitled "Por un Futuro Libre de Combustibles Fósiles (For a Free Future of Fossil Fuels (2018), during the period 2014-2016, a total of 11 European countries and the EU diverted at least 112,000 million euros in the form of subsidies for the production and consumption of fossil fuels in the European continent.

materials they need and develop trade and investment agreements, among others.

In addition to public support in the use of fossil fuels for different purposes, the economic support of the private sector to the use of this type of energy source complicate even further the transition to a cleaner economy in some countries. Banks, insurance companies, pension funds, and hedge funds, among others provide companies with one billion euros each year in loans for projects, bonds, and other financial services associated to the use of fossil fuels in specific economic sectors, which allows them to continue expanding the use of this type of energy source at world level.

Another critical issue under consideration in several countries related to the future use of fossil fuel for electricity generation and other purposes is the potential for safe and effective geological sequestration of CO_2 in deep underground formations.

If fossil fuel is expected to continue to be an essential component of the energy mix in several countries, then it is indispensable that energy intensity will continue to improve during the coming decades. In this case, it is expected that by 2040 many countries might be able to generate 75% more income globally per unit of energy used. This depends on the results obtained of the ongoing structural shifts carried out by several countries moving from high energy-intensive economic activity to less energy-intensive economic situation as well as improving technology (Copenhagen Economics, 2017). It is important to stress that energy intensity decreases as countries grow richer by "changing the economic structure, from energy-intensive manufacturing to less energy-intensive services and by increased efficiency of a range of energy processes, through improved technology, as well as improving capital markets and investment capacity" (Copenhagen Economics, 2017).

However, world energy intensity is expected to continue to fall almost by 20% during the coming years until 2040. A key objective of an energy transition is to reduce energy intensity as much as possible. The comparison of carbon intensity (Kg CO_2/Gj) for coal, oil, and natural gas is the following: coal 90 Kg CO_2/Gj; oil 71 Kg CO_2/Gj, and natural gas 50 Kg CO_2/Gj (Figuera, 2017).

Finally, it is essential to stress the following: according to the calculation made by several experts more than 80% of the world's energy supply comes from fossil fuels, that is, crude oil, natural gas, and coal[33]. Globally, today crude oil provides around 35% of global energy needs, coal more than 25%, and natural gas more than 20%. The remaining 20% is supplied by nuclear energy (6%) and renewable energy sources (14%), among which are solar, wind, hydroelectric power, biomass and other waste that are not usually exchanged in the markets.

In its latest World Energy Outlook 2014 report, the International Energy Agency considers that by 2040 the world's primary energy supply will be divided into four practically equal parts:

- Crude oil with 26% of the total, an expected reduction of around 9%;
- Natural gas with 24% of the total, an expected increase of at least 4%;
- Coal with 24% of the total, an expected reduction of at least 1%;
- Renewable with 17% of the total, an expected increase of 3%, and
- Nuclear, with 7% of the total, an expected increase of 1%.

It is important to single out that the group of industrialized countries consumed more crude oil and natural gas for electricity generation and heating that other countries, while the group of developing countries consumed more coal than others with the same purpose. However, it is expected that by 2030, natural gas will be the primary fuel of the energy mix of the OECD countries, driven by new regulations that limit emissions in the US and other countries in the electricity generation sector. It difficult to predict, in a precise way, how long the world reserves of natural gas will last, but what is clear is that the demand for this non-renewable energy increases from year to year, and it is probably to continue to grow in the

---

[33] It is important to be aware of that fossil fuels are expected to continue to be the dominant form of energy in the supply of global expansion, providing 60% of the additional energy required and representing almost 80% of the total energy supply by 2035 (BP Energy Outlook 2016 edition. Outlook to 2035, 2016).

coming years in comparison with other fossil fuels. Since the first oil crisis, natural gas consumption has suffered a significant increase of more than 220%. Forecasts estimate that this increase in demand will follow the same line and will even triple in certain emerging countries.

On the other hand, and according to the trends included in the World Energy Outlook 2014 report on crude oil demand during the 2012-2040 period[34], there is a vast difference in crude oil demand from one region to another. Thus, for each barrel of crude oil that is left to be consumed in the OECD countries, two more crude oil barrels are consumed by countries outside this organization. In 2040, the Asia and the Pacific region is expected to import two of every three barrels of crude sold in the international market.

According to the document BP Energy Outlook 2016 edition. Outlook to 2035, the changes that it is expected to take place in the energy importance of fossil fuels within the energy mix of many countries around the world it is likely to happen mainly due to the interaction of two factors. These factors are the following:

- The growing participation of renewable energy for electricity generation and heating in many countries, whose contribution to the energy mix is expected to increase from 2,7% in 2015 to 7,9% in 2035;
- The likely lower weight of liquid fuels, which would reduce their participation in the energy mix of several countries from 32,8% in 2015, to 29,6% in 2035. However, the share of natural gas within the energy mix of these countries is expected to increase from 24,2% in 2015 to 25,6% in 2035, but this increase will not be enough to contain the likely decreasing behavior of coal, which will decrease from 29% to 24,7% during the period 2015-2035.

---

[34] According to the World Energy Outlook 2014 report, it is foreseeing that the higher consumption of petroleum in the transport and petrochemical sectors will cause an increase in crude oil demand, which is expected to go from 90 million barrels per day in 2013, to 104 million barrels per day in 2040, although high prices and new energy policies could slow down the growth rate of crude oil consumption as a whole.

Summing up the following can be stated:

- It is expected that fossil fuels will continue to provide, in 2040, most of the energy that many countries need. However, many of these countries will, at the same time, continue to move away from the use of coal for electricity generation and heating to the use of natural gas for the same purpose as much as they can. Coal use in electricity and heat generation could increase by 60% unless countries with highest growth rate all over the world switch to use other energy sources to meet expected growing power demand. In any case, in the immediate future, the world is likely to move towards a new energy mix that will include different sources of energy and will not result in a massive displacement of fossil fuels by other types of energy sources. It is expected, that 85% of the growth in world energy demand will be covered by renewable and natural gas. It is also likely that renewable energies will become the largest source of electricity generation and heating worldwide in 2040. "The world continues to electrify, with around three-quarters of the increase in primary energy absorbed by the power sector" (BP Energy Outlook 2019, 2019);
- It is probable that the use of coal in several countries will be concentrated in the industrial sector in the future, where alternatives to the use of this type of energy source are more challenging to find[35].
- For this reason, the use of coal for steelmaking is likely to increase during the coming years. "However, growth may be lower than growth in steel production, reflecting a trend towards steel processes using natural gas and electricity. Coal use also persists in high-temperature industrial processes as well as residential and commercial heating, but growth is smaller than for other fuels

---

[35] Coal would be increasingly concentrated in steelmaking and industry. This includes as feedstock for steelmaking but also for high-temperature applications (notably, metals and minerals).

(notably, natural gas), reflecting concerns about air quality and other factors" (Copenhagen Economics, 2017);

- By 2040, many countries are expected to reduce to the minimum or will eliminate completed the use of coal within the power sector[36]. Once the current wave of coal-fired power plant projects under construction is over, the flow of new coal-fired power plant projects starting operation slows sharply after 2020. However, it is too soon to count coal out of the energy mix in many countries, particularly in developing countries. With industrial coal use showing a slight increase to 2040, overall global coal consumption is expected to be flat, with declines in China, Europe, and North America offset by rises in India and Southeast Asia (World Energy Outlook 2018, 2018)[37]. It is significant to indicate that in 2017, coal generated 40% of the world's energy. In that year, there were 530 coal-fired power plants under construction all over the world (Beck, 2017);
- Oil use is expected to be concentrated in the coming years in the transport sector and as feedstock for chemicals production. The use of oil for electricity generation and heating will be lower than today in several countries, including those with a high output of this type of energy source.

For this reason, several big oil companies are investing less in oil and more in natural gas, as well as in different renewable energy sources for electricity generation and heating. The perspective of many oil-producing countries is to adopt several measures to speed up the substitution of oil for electricity generation and heating for renewable or any other less polluting alternative energies.

---

36 According to different sources, it is expected that, by 2020, India will displace the US as the second largest consumer of coal in the world, and shortly thereafter do the same with China as the largest importer of this type of energy source. In 2017, the US reached 17% of the world primary energy consumption. The World Energy Outlook 2014 report foresees that in 2040, China, India, Indonesia, and Australia together will represent more than 70% of world coal production, which underscores the importance of Asia in the international market for this type of fossil fuel.

37 Contrary to the opinion of several energy experts, Khalid Al-Falih, Saudi Arabia's energy minister, estimates that the world will still need to invest in oil to the tune of almost US$1 trillion a year for the next 25 years to satisfy expected growth in oil demand all over the world.

- On the other hand, oil use in the transport sector is expected to plummet over the next 25 years, to be replaced using electricity, natural gas, and biofuels;
- Natural gas would continue to be a relevant component of the energy mix of several countries, particularly for electricity generation and heating. Natural gas is expected to surpass coal in 2030, becoming the second-largest fuel in the global energy mix, although by that year natural gas is probably the primary fuel of the energy mix of the OECD countries. The main characteristics of this new energy cycle are:
  a) the increased production of unconventional gas in the US and other countries in recent years, and the discovering of new deposits;
  b) the increase in the share of LNG in world energy trade from 31% in 2008 to 42% in 2035; and
  c) the growing tendency to decouple the prices of natural gas and oil. However, it is expected that by 2040, the global share of natural gas in the power world sector could decline to 10-15%. Despite this decrease, natural gas and oil is likely to continue to meet a significant share of global energy demand in 2040;
- From a geographical perspective, an essential factor for the future of natural gas will be the objectives that are established in terms of carbon reduction, air quality, as well as the access to the supply that the leading emerging markets could adopt, particularly in the case of China[38], and India. On the other hand, climate change, along with the security of natural gas supply remains a constant concern by

[38] In 2013, Xi Jinping came to power and announced drastic changes in the country's growing pollution problems by promoting an energy policy called "war on pollution." China is, according to several sources, the world's largest energy consumer. In 2017, the country consumed 3.132 million ton of oil equivalent or 23% of the world primary energy consumption. As a result of the implementation of the above-mentioned energy policy, many coal-fired powers plants have been closed or replaced with highly efficient and less polluting clean energy alternatives, such as wind, solar, and natural gas. As a result, the share of coal used in China's total energy consumption is expected to decline from around 66% in 2015 to less than 45% by 2035; this means a fall of 21%. These results have led China to lead the shift to natural gas as an energy resource, and it is expected that its fuel consumption will almost double to 11% by 2035.

European and Asian countries. In the specific case of the European region, the main concern is the role of Russian natural gas in satisfying the foreseeable energy needs of several countries within this region, while in Asia it is to ensure the competitiveness of imported LNG compared to other national energy sources such as coal and renewable energy;

- According to the World Energy Outlook 2018 report, "today's flow of new upstream projects appears to be geared to the possibility of an imminent slowdown in fossil fuel demand, but in the New Policies Scenario, this could well lead to a shortfall in supply and a further escalation in prices. The risk of a supply crunch looms largest in oil. The average level of new conventional crude oil project approvals over the last three years is only half the amount necessary to balance the market out to 2025," given the expected demand outlook until that year.

Finally, it is important to single out the following: There is a profound shift in energy consumption from other regions of the world to Asia in across all fuels and technologies, as well as in energy investment[39]. Also, it is important to stress that the international energy trade flows increasingly shift to Asia from across the Middle East, Russia, Canada, Brazil, and the US[40]. Asia's share of global oil and gas trade rises from around 50% of world total today to more than 66% of the world total by 2040; this means an increase of more than 16% in the next two decades. On the other hand, "Asia makes up half of the global growth in natural gas, more than 80% of the increase in oil, and more than 100% of the growth in coal and nuclear" (World Energy Outlook 2018, 2018).

The most critical issue regarding the future use of fossil fuel is the magnitude in which this type of energy source is used today at the world level (around 80% of the total energy needs in both developed and

[39] According to different experts' calculation, to meet the expected demand, it will be necessary to invest in oil and gas exploration and production close to US$0.9 trillion per year until the 2030s,

[40] There is clear competition between natural gas producers in the US, Australia, Qatar, and Russia. These four countries seek to project their energy exports throughout the world, with empathizes in Europe and Asia.

developing countries). For this reason, the transition to a system where fossil fuels were not the majority is not a matter of years, or decades, but of many decades[41]. One of the main reasons that prevent sudden or rapid changes in the composition of the energy mix in many countries is the significant investments needed to start up the energy infrastructures that need to be built to allow the use of new energy sources economically.

Based on what has been said in the above paragraphs, the following can be stated: the life of the planet and the current model of world energy consumption is not sustainable indefinitely and should be changed as soon as possible[42]. On the other hand, the earth has been subjected to global warming, droughts, hurricanes, extinction of species, environmental pollution, increased diseases, alterations in agricultural productivity, among other events that are threatening the survival of human civilization. The trigger of this situation is undoubtedly the use of fossil fuels for electricity generation, heating, and for other purposes in several economic sectors of many countries all over the world.

POSSIBLE ACTIONS TO BE TAKEN

Bearing in mind all that has been said earlier, it is crucial to promote the use of alternative fuels, such as renewable and nuclear energy were possible, for electricity generation and other purposes in many countries. The aim is to reduce the contamination of the environment and the population as much as possible at the world level.

The following are several actions that could be taken by all interested countries to achieve the goal mentioned above:

---

[41] For example, what happened during the period after the oil crises of 1973 (1973-2008): in the 35 years that have passed since the first oil crisis reported in 1973 and until 2008, the contribution of oil to the energy mix went from 46% to 33% in the world as a whole and from 52% to 37% in the OECD countries (IEA, 2010), that is, a reduction of only 13% and 15% respectively.

[42] There is an energy plan adopted by the European Union, called "Third Industrial Revolution," which involves the replacement of 70% of non-renewable energy by non-polluting renewable energy by 2050.

- The significant reduction of financing in the use of fossil fuels for electricity generation and heating, among other purposes by governments and the industry[43]. The subsidies and other facilities granted to producers and consumers of fossil fuels such as oil, coal, and natural gas increase global warming and delay the development of renewable energies and the transition to a cleaner energy economy. This assistance is materialized as subsidies, mostly for consumption but also for production, and have a high economic, environmental and social cost that slow the transition to a low carbon economy. In addition, energy subsidies damage the environment, impose substantial fiscal costs, which need to be financed by some combination of higher public debt, higher tax burdens, crowding out of potentially productive government spending, discourage investments in energy efficiency, renewables, energy infrastructure, and increase the vulnerability of countries to volatile international energy prices (Coady, et al., 2015);
- Adoption of additional energy and saving measures on the one hand, and to improve energy efficiency in the other[44]. The current energy system is extraordinarily inefficient, and, for this reason, a transition cannot be carried out without first reducing energy demand at the world level. Energy companies have agreed to gain an efficiency now crucial in their margins in the cheapest oil era of recent years. While prices were high, energy companies did not worry too much about reducing costs, but now they have learned -by obligation- to be profitable with cheap oil. If the world wants to abandon fossil fuels, the first solution is to reduce energy consumption in many countries;
- The promotion of environmental culture. This type of lifestyle is related to the educational process aimed at awakening in human

---

[43] For example,, in some Latin American and Caribbean countries, subsidies are applied to the consumption of fossil fuels such as gasoline, diesel, liquefied petroleum, and natural gas, among others. During the period 2008 and 2014, the subsidy for gasoline averaged 38% of the total amount of subsidies given annually, diesel 28%, and natural gas 22%.

[44] World average energy efficiency was US$8.617 per ton of oil equivalent in 2017. Between 2007 and 2017, world average energy efficiency grew at an average annual rate of 1,7%.

beings a conscience about the environment. This process tries to promote a change in the relationship between man and the natural environment, guaranteeing the sustainability and quality of the ecosystem for both the current and future generations. The environmental culture can help to improve the quality of life of the living beings of the planet. According to BP Energy Outlook 2019 report, this improvement in living standards is likely to cause energy demand to increase by around a 33%, driven by India, China, and other Asian countries, which together it is expected to account for about 66% of the increase.

For this reason, the adoption of an environmental culture is a responsibility of the civil society, governments, official and academic institutions, the economic sectors with impact on the environment, among others;

- Sustainable economic development. Some of the worst effects on sustainable economic growth "are expected to be felt among agricultural and coastal-dependent livelihoods, indigenous people, children and the elderly, poor labourers, poor urban dwellers in African cities, and people and ecosystems in the Arctic and Small Island Developing States" (Roy, J., et al., 2018);
- Reduction of greenhouse gases by industrial factories, automobiles, and countries all over the world. Governments around the world have been working on the use of clean coal, and have already spent US$10 billion on carbon capture through capital investments. But according to the Global CCS Institute, "the world needs 2.000 carbon capture and storage facilities around the world by 2040 to keep the world under two degrees of warming and avoiding some of the worst consequences of climate change. Currently, the world has 17 large-scale CCS facilities, with four more expected in 2018" (Beck, 2017).
- The rational and conscious use of natural resources. The overexploitation of natural resources, in most cases indiscriminately, has had a high negative impact on the planet and the population. For this reason, not only is essential the individual

practices of the rational use of energy but also the adoption and implementation of environmental policies that encourage the correct use of resources all over the world;

- The increased use of all available renewable energies such as hydro, wind, solar, geothermal, tidal, and biomass. Without a doubt, the development and increased use of renewable energies are essentials to fight climate change, limit its most devastating effects, and reduce the need to use fossil fuels for electricity generation and heating in many countries. The transition to an energy system based on the use of renewable energy sources for electricity generation will also have very positive economic effects. According to the International Renewable Energy Agency (IRENA) 2015 report, doubling the share of renewable energies in the world energy mix to reach 36% in 2030 would imply an additional global growth of 1,1% that year (equivalent to US$1,3 billion), an increase in welfare of 3,7%, and an increase in employment in the sector to more than 24 million people, compared to the current 9,2 million;
- The reforestation: The primary purpose of reforestation is to improve environmental conditions since forests produce most of the oxygen the people breath and help to reduce the greenhouse effect. Besides, they regulate the climate and are part of the natural habitat of many plant and animal species. The importance of reforestation lies in guaranteeing enough oxygen on Earth for the survival of the species. The lack of trees causes the Earth's heat to be suspended in the atmosphere and not be released, due to the high levels of CO_2 and other gases that impede it. The trees are responsible for promoting and filtering CO_2, not only eliminating it from the planet but transforming it into oxygen, an essential element for life;
- Recycling: It is one of the easiest ways to combat global warming since it avoids generating more pollution. Recycling benefits the environment and the economy of a country or region reduces the need for landfills and incineration, saves energy, prevent contamination caused by the extraction and processing of raw materials, and reduces emissions of greenhouse gases that

contribute to global climate change. In other words, conserve natural resources and helps sustain the environment for future generations;

- Compliance with environmental laws and international agreements such as the Paris Agreement on climate change. According to several experts' opinions and projections, the first of the fossil fuels that are expected to be eliminated from the energy mix in many countries is coal, following by crude oil, and natural gas. At the latest in 2050, and according to European energy policy adopted by its member states, Europe must be decarbonized by that year;

CONCLUSION

Human beings have used fossil fuels as their primary source of energy and governments and the industry for their economic growth. One crucial element that should be in the mind of governments and the energy industry during the consideration of energy issues is that fossil fuels were used in the past without measuring the environmental impact they have on the atmosphere. Has been proved that the use fossil fuels in different economic sectors are the main responsible for the greenhouse effect and global warming, and without too much concern about the negative impact on the population, the level of the fossil fuel reserves available, and its rate of consumption.

The indiscriminate exploitation of fossil fuels is the leading cause of the increased in the environmental pollution, provoking the weakening of the ozone layer, altering the greenhouse effect and, subsequently, the appearances of other global environmental problems such as global warming and climate change. For this reason, it can be stated that the main problem of burning fossil fuels is that they release significant quantities of the main gas responsible for the greenhouse effect, such as carbon dioxide CO_2, contributing to the increase in atmospheric temperature and climate change. For this reason, if the world wishes to reduce the negative impact on the environment and to stop climate change, all countries should reduce significantly or to discontinue the use of fossil fuels for electricity generation and heating.

Despite the negative impact on the environment and the population, fossil fuels are and will remain in the future, as a competitive energy source, in comparison with other energy sources such as renewables and nuclear energy. There are several reasons for this. These are the following:

- Fossil fuels are abundant and can be found in many countries all over the world. However, it is a fact that all fossil fuels are finite and are non-renewable, therefore, exist in a limited manner from a physical point of view;
- Fossil fuels technology and distribution network are well established in almost all countries. Fossil fuel technology is the primary source to produce movement or other sources of energy. Extraction and conversion mechanisms such as motors, chimneys or power plants and machines such as cars, boats, and ships, airplanes, heaters, among others, use this type of energy source to function;
- The resources consumed to build the current infrastructure for the production and consumption of all fossil fuels in different economic sectors are enormous and were spent in several years.
- On the other hand, it is a fact that technological transitions are slow and require the order of decades to be carried out (Fouquet, 2010), and the rates of change that are being observed at present at the global level are not high. Only a few countries are reducing the use of fossil fuel for electricity generation and heating significantly and, at the same time, increasing, in a significant manner, the use of renewable for this specific purpose[45];
- The consumption of fossil fuels for electricity generation and heating is more effective than the use of other energy sources, despite its negative impact on the environment and the population;
- Energy density is relatively high.

[45] Costa Rica has plans that, by 2050, all electricity generation in the country will be produced by renewables. The UK and the Netherlands have similar plans. Germany expect that by that year around 95% of its electricity generation will be produced by renewables.

It is essential to continue using fossil fuels for electricity generation and heating, the adoption of additional measures to reduce CO_2 emissions urgently and to cut the negative impact on the environment and the population. There no unique and transparent solution to this situation. The implementation of the actions suggested above will improve the energy mix, increase the use of renewables for electricity generation and heating, and reduce the use of coal and oil for this specific purpose. These measures will also limit CO_2 emissions from the burns of coal in coal-fired power plants, increase the use of renewables and nuclear energy, where possible, and improve energy efficiency in electricity generation, among others.

The development and increased use of renewable and nuclear power, where possible, is essential to fight climate change, limit its most devastating effects, and reduce the need to use fossil fuels for electricity generation and heating in many countries. The transition to an energy system based on the use of renewable energy sources and nuclear energy, where possible, for electricity generation and heating will also have very positive economic effects at the world level.

Summing up the following can be stated regarding the future use of fossil fuels for electricity generation, heating, and other purposes: after more than two centuries of the predominance of fossil fuels, the idea that throughout the 21st century there will be a gradual substitution of renewable energy sources is becoming a reality. The most radical proposals in this regard propose 100% renewable electric power generation, but this is something that can be reached by a minimal number of countries. In most countries, fossil fuels will continue to provide most of the energy at the world level during the period 2020-2050 even in the case of a significant increase in the use of renewables. However, the pattern of use will change considerably because many countries will move away from the use of coal and crude oil for electricity generation and heating giving high preference to the use of natural gas for these specific purposes. The use of coal will be concentrated in the industry, where alternatives to the use of this energy source are more challenging to find.

The gradual depletion of fossil fuel resources is a widely recognized ongoing process. There are several studies on the prospects of exhaustion of various fossil fuels, but few of them offer a global perspective of all energy sources and their possible technological substitutes. According to the Copenhagen Economics report (2017), the outlooks in the use of fossil fuels are the following:

Coal

- The consumption of coal is expected to be increasingly concentrated in steelmaking and industry. By 2040, it is foreseen that most of thermal coal use in the power sector would be phased out, even if carbon capture and sequestration[46] could be appropriately used. Several countries have already decided to shut down old and ineffective coal-fired power plants during the coming decades, or are considering a significant reduction in the use of coal for electricity generation and heating;

Crude Oil

- The peak of conventional oil was reached around the year 2006, according to Murray and King (2012). To avoid rapid depletion of the current reserves of crude oil, and to reduce its negative impact on the environment and the population, the use of this type of energy source is expected to be concentrated in the transport sector and as feedstock for chemicals production. A significant increase in the use of electric car and buses is projected to be reported during the coming years to satisfy future passenger transport needs in many countries. It is also anticipated an increase in the efficiency in the use of energy sources in the transport sector at world level;

---

[46]Carbon capture and sequestration technology compresses and transports carbon dioxide fired before or after fuel combustion, through geological and oceanic pipelines, to inject it into the Earth. The purpose in the use of this technology is to reduce, as much as possible, the number of toxins that are thrown into the atmosphere diminishing the negative impact on the environment and the population provoked by the burned of fossil fuels.

Natural gas

- The consumption of natural gas is expected to continue to be a relevant component of the energy mix in several countries during the coming decades. In the power sector, this type of energy source has half the emissions intensity of coal and, for this reason, can be considered as a clean energy source for electricity generation and heating.

In an ideal scenario, the world would see, during the coming decades, a significant advance in innovation and development of renewable energies[47], an increase in energy efficiency, a better energy accumulation systems, the use of more smart grids, and growth in the use of capture and storage of carbon technology, especially for natural gas-fired power plants.

Before 2050, radical policies aiming at a clean future should be consolidated in many countries all over the world. These policies will support the transition from oil vehicles to electric vehicles with a huge technological challenge in the air and maritime transport, and an increasing application of taxes on fossil fuels use for electricity generation and heating. In the coming years, more and more countries will prohibit the use of coal-fired power plants for electricity generation and heating.

References

Beck, Madelyn (2017); Cleaning Up Coal: Folly or The Future? *Inside Energy;* 2017.

BP Statistical Review of World Energy 2016; 2016.

BP Energy Outlook 2016 edition. *Outlook to 2035*; 2016.

[47]It is important to be aware that renewable energies, especially wind and solar, appear as the main sources of energy capable of replacing the decline of fossil fuels (Smil, 2010). However, some recent studies show that their maximum exploitation potentials at the global level can be significantly lower than the current consumption of energy from fossil fuels (de Castro et al., 2011, 2013a, and 2013b).

BP Statistical Review of World Energy 2017; 2017.

BP Statistical Review of World Energy 2018; June 2018.

BP Energy Outlook 2019 Edition; 2019.

Canada's Energy Future 2016. *Energy Supply and Demand Projections to 2040* (2016); National Energy Board: An Energy Market Assessment; 2016.

Castro, Noiraly (2018); *The Power of Carbon*; Steemstem; 2018.

Clews, R. J. (2016); Fundamentals of the Petroleum Industry, in Project Finance of the International Petroleum Industry; *ScienceDirect*; 2016.

Coady, David, Parry, Ian, Sears Loiuis and Shang Baoping (2015); *How Large Are Global Energy Subsidies?* WP/15/105; IMF; 2015.

De Castro, C., Mediavilla, M., Miguel, L. J., Frechoso, F. (2011); Global wind power potential: Physical and technological limits; *Energy Policy* 39, 6677–6682; doi:10.1016/j.enpol.2011.06.027; 2011.

De Castro, C., Carpintero, Ó., Frechoso, F., Mediavilla, M., de Miguel, L. J. (2013a); A top-down approach to assess physical and ecological limits of biofuels; *Energy*. doi:10.1016/j.energy.2013.10.049; 2013.

De Castro, C., Mediavilla, M., Miguel, L. J., Frechoso, F. (2013b); Global solar electric potential: A review of their technical and sustainable limits; *Renewable and Sustainable Energy Reviews* 28, 824–835; doi:10.1016/j.rser.2013.08.040; 2013.

Figuera, Luis (2017); The Future of Fossil Fuels (Part 1 and 2); *Linkedin*; 2017.

Fossil fuel (2019); *Wikipedia*; 2019.

Fouquet, R. (2010); The slow search for solutions: Lessons from historical energy transitions by sector and service; *Energy Policy* 38, 6586–6596; doi:10.1016/j.enpol.2010.06.029; 2010.

Hudson Teslik, L. and Sparks, K. (2009); The Future of Fossil Fuels; Originally published in the *Britannica Book of the Year*; 2009.

International Energy Outlook 2016; *US Energy Information Administration (EIA)*; May 2016. Online at: https://www.eia.gov/outlooks/archive/ieo16/; 2016.

International Energy Outlook 2016 with Projection to 2040; *US Energy Information Administration, Office of Energy Analysis, US Department of Energy, IEO 2016*; 2016.

International Energy Agency (2010); *Key World Energy Statistics 2010*; OCDE/IEA; Paris; 2010.

International Energy Agency (2015); *National Geographic Magazine in its special issue of Climate Change (November 2015), 10 arguments in favor of renewable energies - Sustainability for all*; 2015.

Moriana, Laura (2018); Loss of Biodiversity; Causes and Consequences; *Green Ecology;* 2018.

Murray, J., and King, D. (2012); Climate policy: Oil's tipping point has passed; *Nature* 481, 433–435; doi: 10.1038/481433a; 2012.

Oil (2017); *Institute for Energy Research (IER)*; 2017.

OPEC database; 2017.

Por un Futuro Libre de Combustibles Fósiles [*For a Future Free of Fossil Fuels*] (2018); Amigos de la Tierra; 2018.

Quiroga Martínez, Borja (2017); *La Electricidad Frente al Combustible* [*Electricity vs. Fuel*]; Twenergy; 2017.

Ritchie Hannah, and Roser Max (2018); *Fossil Fuels*. Published online at OurWorldInData.org. Retrieved from: https://ourworldindata.org/fossil-fuels [Online Resource]; 2018.

Reid, Gerard (2018); *The Future of Coal*; 2018.

Roy, J., Tschakert, P., Waisman, H., Abdul Halim, S., Antwi-Agyei, P., Dasgupta, P., Hayward, B., Kanninen, M., Liverman, D.,Okereke, C., Pihno, P. F., Riahi, K., Suarez Rodriguez, A. G., (2018); Sustainable development, poverty eradication and reducing inequalities. In: *Global warming of 1.5°C. An IPCC Special Report on the impacts of global warming of 1.5°C above pre-industrial levels and related global greenhouse gas emission pathways, in the context of strengthening the global response to the threat of climate change, sustainable development, and efforts to eradicate poverty* [V. Masson-Delmotte, P. Zhai, H. O. Pörtner, D. Roberts, J. Skea, P. R. Shukla, A. Pirani, W. Moufouma-Okia, C. Péan, R. Pidcock, S. Connors, R. B. R. Matthews,

Y. Chen, X. Zhou, M. I. Gomis, E. Lonnoy, T. Maycock, M. Tignor, T. Waterfield (eds.)]. In Press; 2018.

Smil, V. (2010); *Energy Transitions: History, Requirements, Prospects; Praeger*; 2010.

The Future of Fossil Fuels: *How to Steer Fossil Fuel Use in a Transition to a Low-Carbon Energy System* (2017; prepared by Copenhagen Economics for the Energy Transitions Commission in January 2017; 2017.

The Truth About Oil: Pros and Cons of Fossil Fuels (2018); *Recond Oil;* 2018.

White Paper on the Future of Europe. Reflections and Scenarios for the EU 27 by 2015 (2017); *European Commission COM* (2017) 2025; 2017.

World Energy Resources 2016; *World Energy Council*; 2016.

World Energy Outlook 2014 (2014); *International Energy Agency*; 2014.

World Energy Outlook 2018 (2018); *International Energy Agency*; 2018.

In: Advances in Energy Research. Volume 31 ISBN: 978-1-53616-271-4
Editor: Morena J. Acosta © 2019 Nova Science Publishers, Inc.

Chapter 2

Ti Microholes Potential for Fossil Fuels Efficient Application Punched by WC/Co Micropunch

Kelvii Wei Guo*

Department of Mechanical and Biomedical Engineering,
City University of Hong Kong, Hong Kong

Abstract

The use of microholes is a potential approach to enhance fossil fuels efficient applications in the various energy fields, especially for thermal power plants.

Ascribed to eco environmental competition, micropunching is extremely suitable for the mass production of micro features with friendly eco effects. Therefore, the morphology variation of micropunch and microhole during the punching with WC/Co micropunches was investigated by scanning electron microscopy (SEM), energy dispersive spectrometer (EDS) and confocal laser. The results reveal that due to the optimal joint-contribution of WC and Co, the formed microholes satisfy with the practical requirements in the quasistable period. Moreover, the

* Corresponding Author's E-mail: kelviiguo@yahoo.com.

serious wear of micropunch occurs with the wear loss both of Co and WC when the punching number exceeds 1525. With the punching numbers further increment, the dominant factors of the wear loss would mainly rely on the easily peeled off WC due to the serious loss of Co. In addition, the microholes can be adequately processed after about 30 min by natural sand grains. The quality of the hole would decrease with further increase in processing time and sand accumulation becomes severe.

Keywords: microhole, thermal power plants, WC/Co, micropunch, morphology

1. INTRODUCTION

It is well known that the thermal power plants can be classified by the source of the energy used to generate the steam that it is expanded in the turbine to produce electricity, which are listed as follows classified by heat source: (1) Heat sources for fossil fuel power stations: (i) a steam turbine generator; (ii) the natural gas-fired plants. They may use a combustion turbine. To date, a coal-fired power station produces heat by burning coal in a steam boiler. The steam drives a steam turbine coupled to a generator producing electricity. It is well known that the fossil fuel power stations are still currently the dominating plants for generation of energy production around the world. (2) A nuclear power plant is a steam turbine plant where the steam is generated by a nuclear reactor. Up to now, about 10% of all electric generation worldwide is produced by nuclear power plants. It should be noted that after the Fukushima disaster, all the plants are scheduled to be shut down in a few years, and the rest of the world is very hesitant with respect to what role nuclear energy should have in future energy planning. (3) Geothermal power plants use steam extracted from hot underground rocks. This kind of energy plant only contributes a minor energy output compared with the global demand. (4) Biomass-fueled power plants are fueled with wood pellets, wood chips, straw, and waste from the agricultural industry such as sugar cane and nut scale, municipal solid waste, landfill methane, or other forms of bio gas. (5) In integrated steel mills, blast furnace exhaust gas is a low-cost, although low energy-density, fuel. This recycled gas is available to produce about 60% of the total electricity consumption. (6) Waste heat from industrial processes is occasionally concentrated enough to use for power generation, usually in a steam boiler and turbine. (7) Solar thermal electric plants use sunlight to boil water and produce steam, which turns the generator. (8) IGCC (integrated gasification combined cycle) with carbon capture and storage (CCS) technology allows coal to be used to generate power as cleanly as natural gas [1, 2].

In most of the countries, their thermal power plants are playing an important role in the energy production. Therefore, the research work should be put enough attention towards the optimization of these power plants. In the developing countries, energy supplies are less secure because of its costlier price. Indeed, there is a need to reconsider the lowest-cost energy options and the relevant techniques. Since, from an energy performance point of view, the first law analysis has been found to be insufficient. So, in thermodynamic analysis of various thermal processes and plant systems, exergy analysis is getting its own importance. It is well known that the total conversion of heat into work is not possible.

Consequently, that part which is available for conversion is termed as exergy. It is a property associated with the state of system and environment, nowadays a useful tool to differentiate between internal irreversibility and energy losses in a process [2]. The performance of thermal power plants can be evaluated through energetic performance criteria, which are electrical power and thermal efficiency. In recent decades, the exergy analysis has been found as a useful method in the design, evaluation, optimization and improvement of thermal power plants [3, 4, 5]. Exergy analysis helps in finding the losses taking place in a system. By this method, energy conversion at different points, various component efficiencies and points of largest losses are easily obtainable and hence it helps in taking necessary action to decrease them [6, 7].

Some researchers have contributed review papers on exergy analysis, which help young researchers to get in touch with the previous problems [8]. In power plants, insights have been provided into various energy and exergy efficiencies which are helpful for design engineers [9, 10] As a result, the improvement in the thermal performance of power generation units and consuming devices can be achieved significantly by combining exergy analyses with the related techniques upgraded.

In recent years, micro technology has become one of the key disciplines with a significant effect on the development of new products and production technologies [11-14].

As is known to all, microtechnology describes the technological approach, directed to the miniaturization of components and systems, down to micrometer scale. Microtechnological components, such as distributed holes, bear the potential to provide further functionality, e.g., enhancing fluid flow and heat exchange within thermal power plants, especially for the turbine blades [15-18].

Up to now, the ever increasing demand for smaller, higher-quality and lower-priced products from almost all fields of industry, household equipment and entertainment electronics involves the optimisation of already existing and the development of new manufacturing methods which are tailor-made for the micro systems technique with higher precision [19, 20].

However, these kinds of micro-devices are mainly fabricated by micro-machining technique, and the fabrication technology with stable and low cost is one of the important issues [11, 12, 14, 19].

Therefore, the microholes formed by micropunching at low cost and in large quantities, applied for thermal power plants application and micro-parts fabrication have been researched. This research aims at investigating the wear characteristics of micropunches (150 μm in diameter) and the morphology variation of microholes formed by punching pure titanium (Ti) in various processing periods to overcome current problems in the micro-metal-forming technology. In the long run, this research can lead to making microholes distributed in the thermal power materials (both non-metals and metals) at low cost and in large quantities with the eco-friendly technique.

2. Experimental Materials and Procedures

2.1. Experimental Material

Micropunch with 75% volume fraction WC particle and 25% volume fraction Co particle of 50 μm mean size, 150 μm in diameter, is shown in Figure 1. Figure 2 shows the surface texture of micropunch. Pure titanium sheet with 200 μm in thickness was used as the substrate.

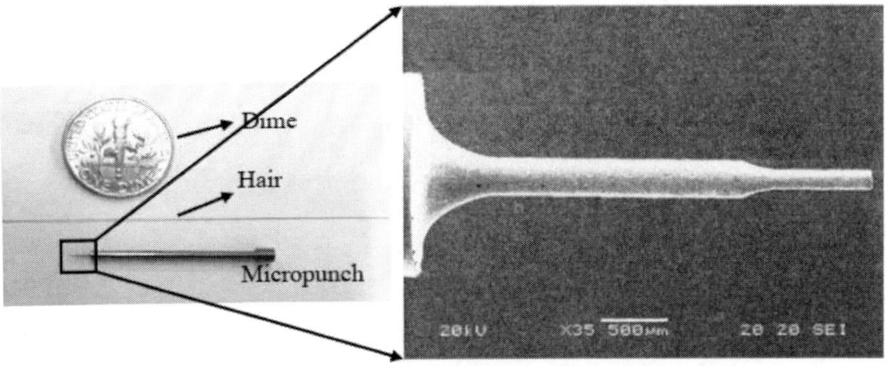

Figure 1. Profile of micropunch.

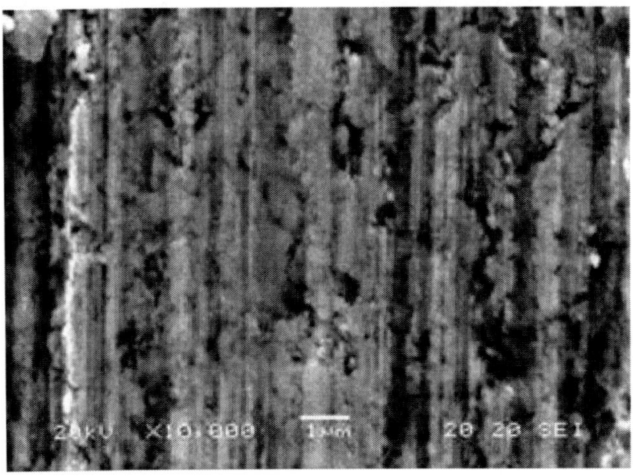

Figure 2. Surface texture of micropunch.

2.2. Experimental Procedures

In order to clean the contaminants in the prepared pure titanium sheet, it was carefully washed by acetone and pure ethyl alcohol before putting into the microdie. After that, the microprocessing machine MP50 (Made in Japan) was taken to punch the titanium sheet with 20 pulses per minute, and feedrate of 2 mm.

The wear of the micropunch and the variation of the morphology of punched microholes in different processing periods were investigated using confocal microscopy, scanning electron microscopy (SEM), energy dispersive spectrometer (EDS) and confocal laser.

3. RESULTS AND DISCUSSION

3.1. Initial Wear Characteristics of Micropunch

The relationship between the wear loss of micropunch and punching numbers is shown in Figure 3. It illustrates that the weight of micropunch (each for 5 times) has an obvious decrease with the increment of punching number in the initial. Its corresponding surface texture is expressed in Figure 4. It depicts that the particles distributed more uniformly than that of parent material (cf. Figures 2 and 4).

According to Figures 3 and 4, it shows that the wear of micropunch in the initial increases significantly, and WC particles cannot be easily observed. Consequently, the dominant factor of the wear loss in the initial period is mainly due to Co. The morphology of the formed microhole is expressed in Figure 5. Some substrate debris is distributed sparsely in the backside as shown in Figure 5b, and its EDS results are illustrated in Figure 6.

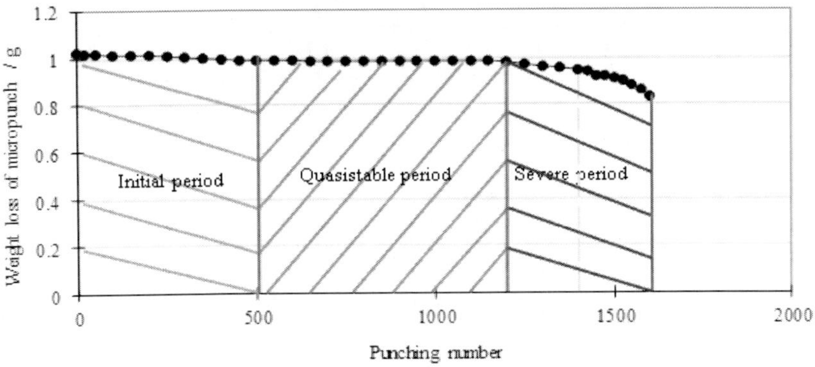

Figure 3. Relationship between wear loss of micropunch and punching numbers.

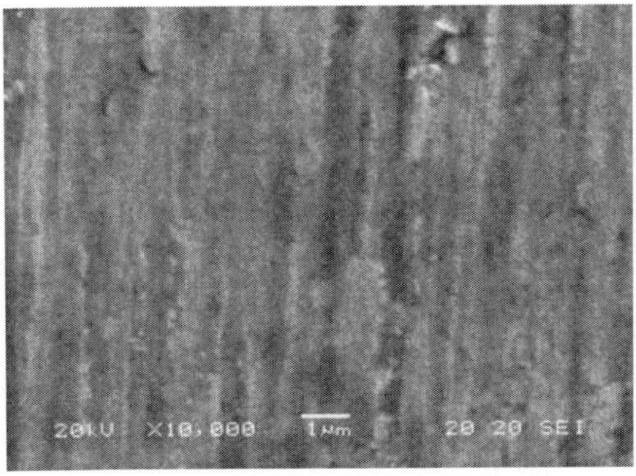

Figure 4. Surface texture of micropunch in the initial condition (Punching number lower than 500).

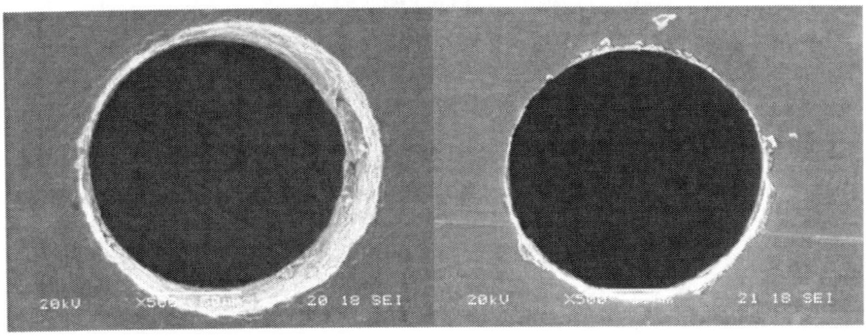

(a) Front side (b) Back side

Figure 5. Morphology of microhole.

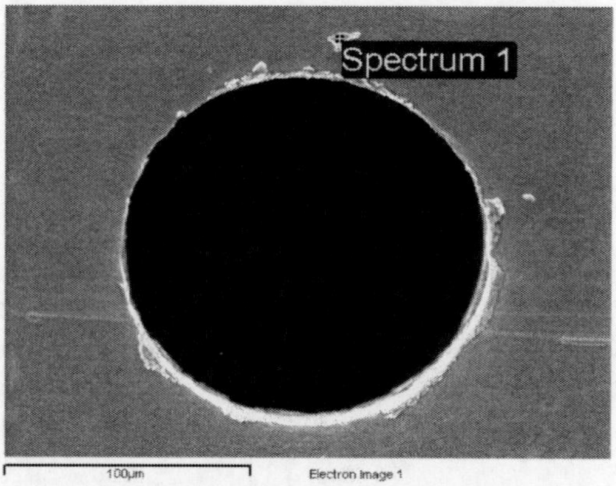

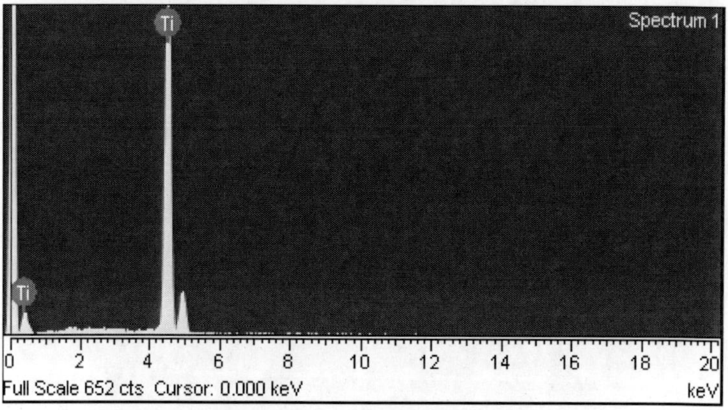

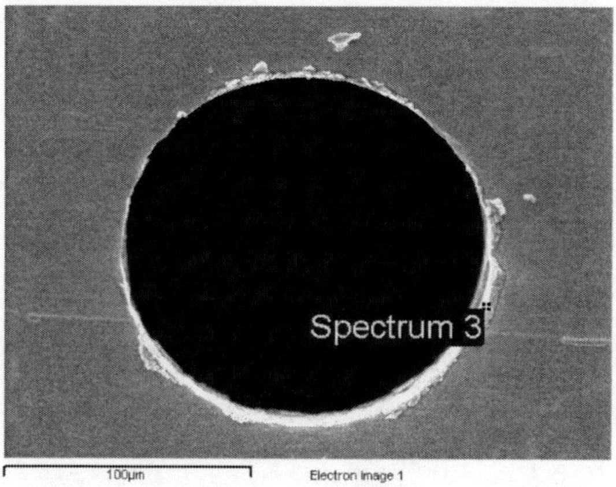

Figure 6. (Continued).

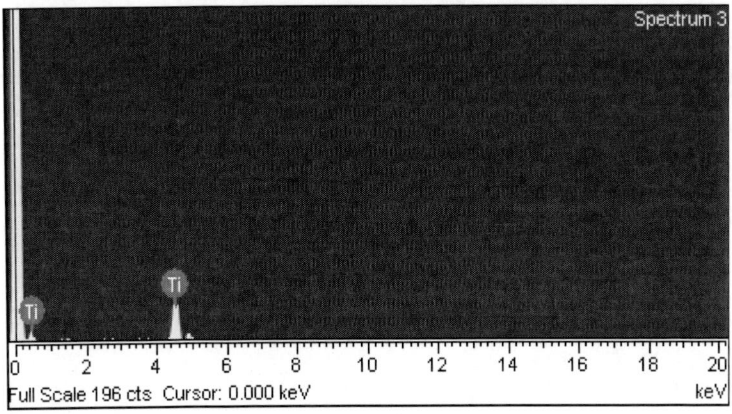

Figure 6. EDS results of debris in the backside.

3.2. Quasistable Wear Characteristics of Micropunch

With the increment of punching numbers, the phenomena of the initial distinct wear of micropunch disappears, the wear loss of micropunch becomes relatively stable with a little variation as depicted in Figure 3, particularly for punching number from 500 to 1200. The related surface texture of micropunch is shown in Figure 7 correspondingly.

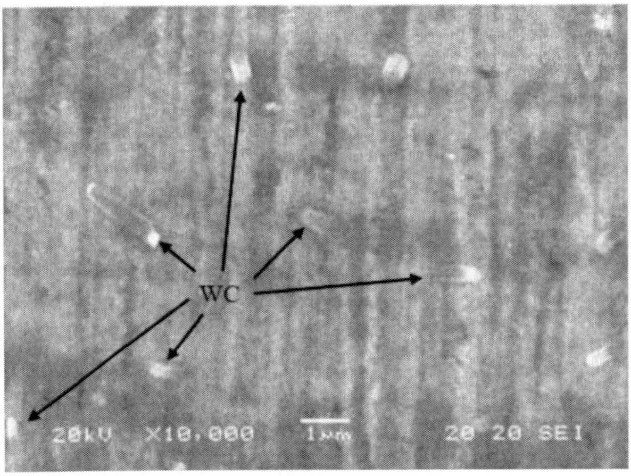

Figure 7. Surface texture of micropunch with punching number from 500 to 1200.

It expresses small pieces of WC particles are observed on the surface. Meanwhile, WC particles distribute uniformly on the surface of micropunch.

The morphologies of the formed microhole are shown in Figure 8. It expresses that although some substrate debris still distribute in the backside (Figure 8b), the quality of the quasistable-period-microhole is obviously superior to the initial period (compared Figure 5 with Figure 8). It illustrates that because of the joint-contribution of WC and Co, the wear loss of micropunch in the quasistable period is little.

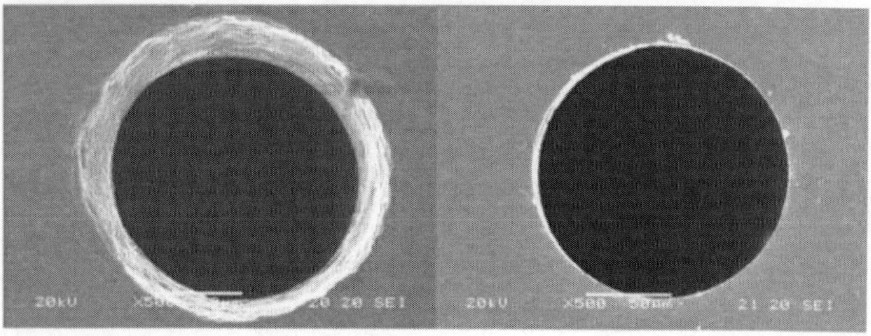

(a) Front side (b) Back side

Figure 8. Morphology of microhole.

3.3. Severe Wear Characteristics of Micropunch

The surface texture of micropunch with punching number over 1525 is shown in Figure 9. It shows that large amounts of WC particles distributes on the micropunch surface because of serious wear loss of Co, which matches well with Figure 3. As a result, the wear of micropunch increases distinctly (Figure 3). Furthermore, with the punching numbers increasing further, the dominant factor of the wear loss would mainly rely on WC (which is easily peeled off during micropunching) as shown in Figure 9.

Figure 9. Surface texture of micropunch in the severe wear condition.

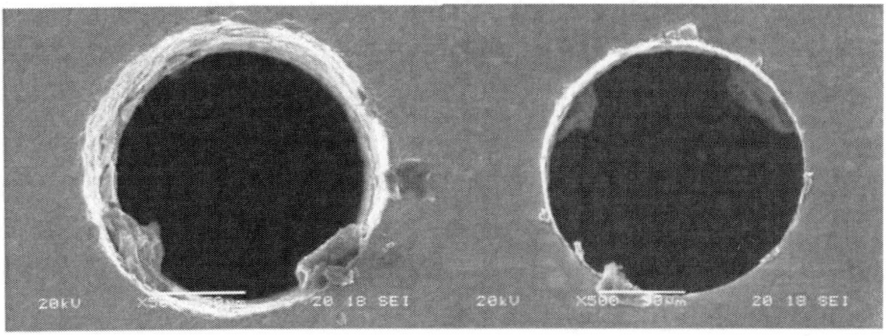

(a) Front side (b) Back side

Figure 10. Morphology of microhole in the severe wear condition.

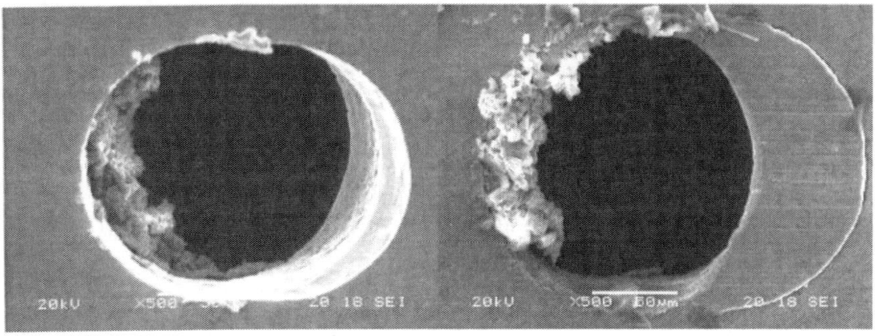

(a) Front side (b) Back side

Figure 11. Morphology of microhole in the intensively severe wear condition.

Meanwhile, the quality of the machined microhole decreases distinctly (cf. Figs. 5, 8, 10 and 11). It illustrates that a large amount of micropunch materials are peeled off and adhered to the Ti substrate. At the same time, the substrate material could not be punched successfully and it would stick to the substrate. Also, with the effect of feedrate, the unsuccessful removed materials would be sheared and formed the larger debris afterwards as shown in Figure 11.

3.4. Profile of Punched Microholes

The diameter of the punched microhole was measured by LEXT confocal laser-OLS3000 as shown in Figure 12. Its corresponding results (each for 5 times) are shown in Figure 13. Compared with Figure 3, it indicates that the diameter of microhole changes in the different micropunching periods. At the beginning of the micropunching, the obvious decrement of the diameter of microhole accompanies with the punching number increasing, which surely agrees well with the results expressed in Figure 3. The attractive results are addressed with the relatively stable diameter of microhole when the micropunching number varies from 500 to 1200. When the punching number increases further, the obvious decrement of the diameter of microhole becomes more and more clear on account of the serious loss of cobalt (Co), especially for the punching number over 1525 as depicted in Figure 13. As a result, in the severe wear period of micropunching, due to the serious loss of bonding material -cobalt, the wear of micropunch mostly relies on the easily peeled off WC particles. Simultaneously, because the temperature of micropunch increases with the increment of the punching number, WC particles are more easily detached off from micropunch. Consequently, the wear loss of micropunch becomes more intensive. Furthermore, if the pulses per minute increases, the wear of micropunch will be drastically enhanced. Therefore, in order to improve the quality of micropunch, the microstructure of micropunch should be considered further with the composition of micropunch, especially for the distribution of WC and Co particles.

It should be noted that although the micropunch is in the so-called severe period, the diameter of the punched microhole is still stable to some degree, especially for the punching number from 1200 to 1400, as shown in Figure 13 highlighted with purple dash line, and its relevant weight loss is tiny as shown in Figure 3 highlighted with yellow dash line though the wear loss of this period is serious.

Figure 12. Profile of microhole punched by micropunch measured by OLS3000.

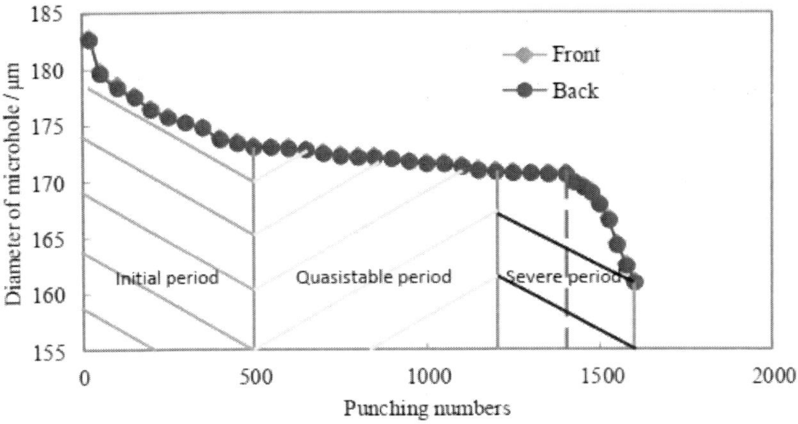

Figure 13. Relationship between diameter of microhole and punching number.

4. Post-Treatment for Punched Microholes by Natural Sand Grains

Features of the punched microholes (such as Figures 5 and 8) include debris (for example, shear marks and burrs). Because of the combination of shear and ductile fracture in the micropunching, the debris produced. Moreover, the property of the processed materials is also one of the crucially effective influence factors on the final features of the punched microhole. In the meantime, the uniformity of the clearance can affect the microhole features during the micropunching (cf. Figures 5, 8, 10 and 11). The debris can hinder the normal functionality of micro features and prevent the proper assembly of micro components to form micro systems. Therefore, the punched microholes should be post treated to improve its finishing.

Figure 14 shows the post treated microhole realized with the agitation of abrasive grains through planetary stirring, where the natural sand was used as the abrasive due to its environmental friendliness. It expresses the results of a 20 min processing microhole. The hole is basically free from loosely attached debris. Burrs at the back side edge are further lowered. Those flattened fragments are slimmer than before. More aggregation of the microscopic grains is observed, both the fragment regions and the edge regions (Figure 14a, insets). Their sizes range from the micron to sub-micron. Continued aggregation might possibly be on account of van der Waals or inter-molecular forces.

With further increment of the processing time, the quality of the microhole continues to improve. Figure 15 reveals that the burrs in the back side edge are no longer existed, except for some small isolated pieces (Figure 15b, inset). Burrs are protrusions from the edge. Their bonding with the substrate is much stronger than that of the re-attached debris. Tiny burrs are particularly hard to be removed. It is well known that the impact energy from a grain of sand depends on the grain size besides its velocity. Those burrs could not be effectively cleaned by the impact from smaller grains. Yet, the chance that tiny burrs are impacted by the larger sand grains is relatively low.

The fragment terraces on the front side become very thin. As a result, they are noticeable mainly under higher magnification (Figure 15a, right inset). In addition, more grains are gathered around the edge of the microhole as well as stuck to the inner wall of the microhole through attraction forces (Figure 15a, left inset).

The quality of the punched microholes cannot improve with further processing. Spontaneously, more sand aggregates after 40 min (Figure 16). The attachment of sand grains to the edge regions and inner wall could be visible from the back side as well as from the front side.

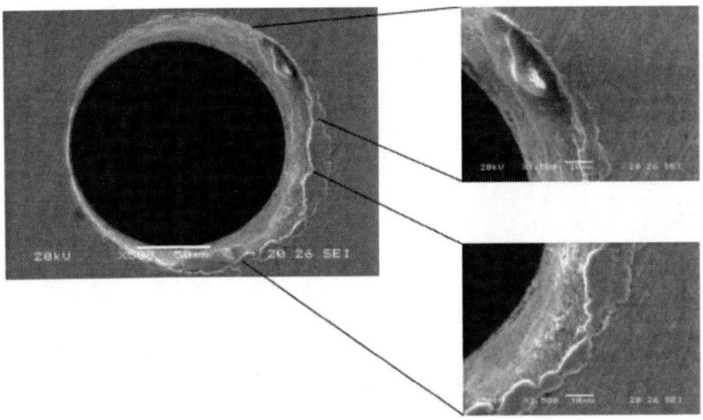

(a) Front side

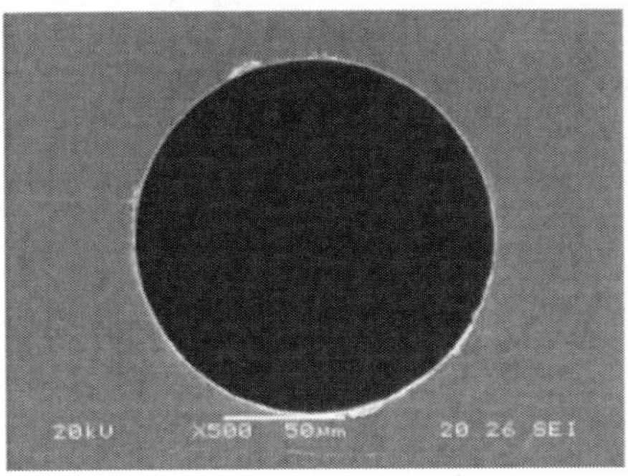

(b) Back side

Figure 14. Morphology of a microhole after 20 min processing.

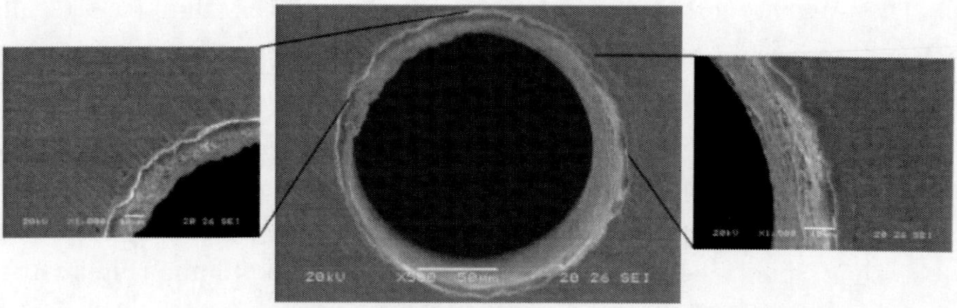

(a) Front side

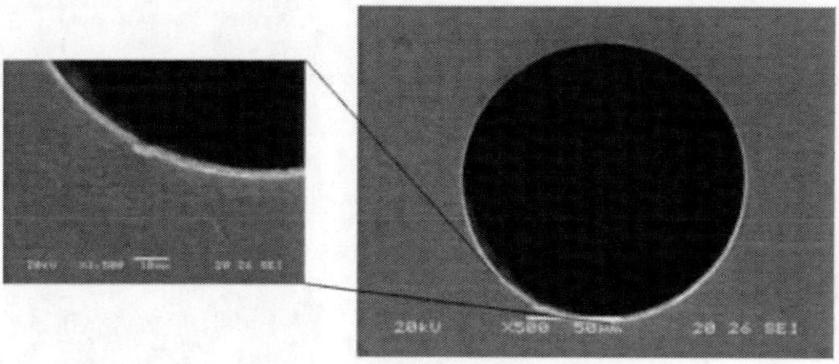

(b) Back side

Figure 15. Morphology of a microhole after 30 min processing.

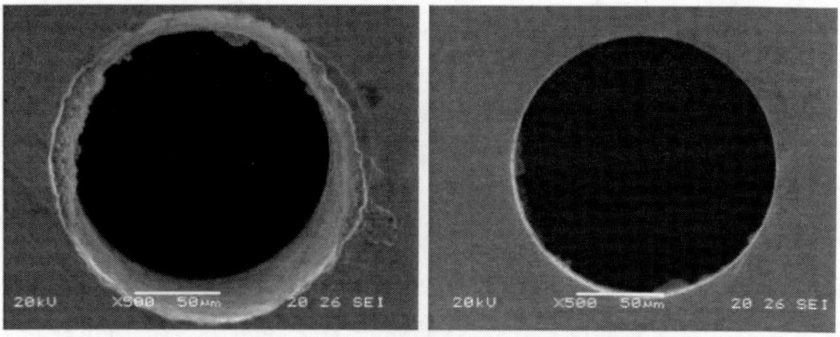

(a) Front side (b) Back side

Figure 16. Morphology of a microhole after 40 min processing.

A band of jagged foil material is detected around the front surface near the edge of the microhole. It is mainly produced as substrate material breaks from the punch when it is pulled back from the hole. It cannot be clearly detected before because (1) the feature size is relatively small compared with those fragments detected at earlier times; and (2) parts of the band are rolled back against the surface of the substrate material. Apparently sand abrasion could only reduce the width of the band to a few microns. Occasionally, parts of the band are bent toward the microhole. On account of the support of sand gathered underneath, it is unlikely that their posture could be reversed in subsequent processing.

CONCLUSION

The wear characteristic of the WC/Co micropunch used for micromachining microhole potential for thermal power plants application had been researched. It shows that the wear of micropunch increases significantly in the initial and the dominant factor of the wear loss mainly relies on Co. With the punching number increasing, the quasistable wear of WC/Co micropunch varies with a little wear loss. However, when the punching number exceeds 1525, the serious wear loss of Co and WC of micropunch takes place. Moreover, with the increment of the punching numbers further, the dominant factor of the wear loss would mostly rely on the easily peeled off WC. Meanwhile, the quality of the punched microholes decreases intensively. In addition, the punched microholes can be adequately processed after about 30 min by natural sand grains. The quality of the hole would decrease with further increase in processing time and sand accumulation becomes severe and no further improvement is observed. Meanwhile, the diameter of the front side of the punched microhole is almost same as that of the back side even in the severe period, which implies the punched microhole can be directly used for improving efficient applications of fossil fuels as a beneficially effective method to enhance fluid flow and heat exchange for the turbines in the power plants.

REFERENCES

[1] Bogmansa, C. W. J., Dijkemab, G. P. J. and van Vliet, M. T. H. (2017). Adaptation of thermal power plants: The (ir) relevance of climate (change) information. *Energy Economics*, 62: 1-18.

[2] Utlu, Z. and Hepbasli, A. (2007). A review on analyzing and evaluating the energy utilization efficiency of countries. *Renewable & Sustainable Energy Reviews*, 11(1): 1-29.

[3] Wang, L. G., Voll, P., Lampe, M., Yang, Y. P. and Bardow, A. (2015). Superstructure-free synthesis and optimization of thermal power plants. *Energy*, 91: 700-711.

[4] Alobaid, F., Mertens, N., Starkloff, R., Lanz, T., Heinze, C. and Epple, B. (2016). Progress in dynamic simulation of thermal power plants. *Progress in Energy and Combustion Science*, 59: 79-162.

[5] Gonzalez-Salazara, M. A., Kirstena, T. and Prchlik, L. (2018). Review of the operational flexibility and emissions of gas- and coal-fired power plants in a future with growing renewables. *Renewable and Sustainable Energy Reviews*, 82: 1497-1513.

[6] Dincer, I. Y. and Cengel, Y. (2001). Energy, entropy concepts and their roles in thermal engineering, *Entropy*, 3(3): 116-149.

[7] Rosen, M. A. and Bulucea, C. A. (2009). Using exergy to understand and improve the efficiency of electrical power technologies. *Entropy*, 11(4): 820-835.

[8] Sciubba, E., and Wall, G. (2007). A brief commented history of exergy from the beginnings to 2004. *International Journal of Thermodynamics*, 10(1): 1-26.

[9] Kanoglu, M., Dincer, I. and Rosen, M. A. (2007). Understanding energy and exergy efficiencies for improved energy management in power plants. *Energy Policy*, 35(7): 3967-3978.

[10] Verkhivker, G. P. and Kosoy, B. V. (2001). On the exergy analysis of power plants. *Energy Conversion and Management*, 42(18): 2053-2059.

[11] Madou, M. J. (2004). Microfabrication challenge. *Analytical and Bioanalytical Chemistry*, 378: 11-14.

[12] Mark, J. J. (2006). *Microfabrication and nanomanufacturing*. Boca Raton FL CRC/Taylor & Francis.

[13] Christophe, G. (2003). Microsystems engineering: metrology and inspection III. *SPIE.*

[14] Fecht, H. J. and Werner, M. (2004). *The nano-micro interface: bridging the micro and nano worlds*. Weinheim Wiley-VCH.

[15] Horn, A., Weichenhain, R., Albrecht, S., Kreutz, E. W., Michel, J., Niessen, M., Kostrykin, V., Schulz, W., Etzkorn, A., Bobzin, K., Lugscheider, E. and Poprawe, R. (2000). Microholes in zirconia coated Ni-superalloys for transpiration cooling of turbine blades. *SPIE*, 4065: 218-226.

[16] Lugscheider, E., Bobzin, K., Moes, M., Lackner, K., Poprawe, R., Kreutz, E. W. and Willach, J. (2005). Laser drilled microholes in zirconia coated surfaces using two variants to implement the effusion cooling of first stage turbine blades. *Advanced Engineering Materials*, 7(3): 145-152.

[17] Kaur, N., Kumar, M., Sharma, S. K., Kim, D. Y., Kumar, S., Chavan, N. M., Joshi, S. V., Singh, N. and Singh, H. (2015). Study of mechanical properties and high temperature oxidation behavior of a novel cold-spray Ni-20Cr coating on boiler steels. *Applied Surface Science*, 328: 13-25.

[18] Moon, H. W., Yoon, Y. J., Park, J. H., Myung, B. S. and Kim, D. E. (2016). Dynamic wetting and boiling characteristics on micro-structured and micro/nano hierarchically structured surfaces. *Experimental Thermal and Fluid Science*, 74: 19-26.

[19] Knobloch, H. and Kaminorz, Y. (2004). *MicroNano Integration*. Springer-Verlag Berlin Heidlberg.

[20] Mohamed, G. H. (2006). *The MEMS handbook*. Boca Raton FL CRC/Taylor & Francis.

In: Advances in Energy Research. Volume 31 ISBN: 978-1-53616-271-4
Editor: Morena J. Acosta © 2019 Nova Science Publishers, Inc.

Chapter 3

REVIEW ON THE DEVELOPMENTS IN PROTON EXCHANGE MEMBRANE DIRECT ETHANOL FUEL CELLS (DEFCS)

C. S. Moreira, V. B. Oliveira **[*]**
and A. M. F. R. Pinto **[†]**
Transport Phenomena Research Center (CEFT),
Department of Chemical Engineering, Faculty of Engineering,
University of Porto, Porto, Portugal

ABSTRACT

Fuel cells are being extensively studied due to their potential as an alternative energy supply technology for a wide range of applications. Among the different types of fuel cells available, direct fuel cells (DFCs) are promising ones to provide energy for electronic devices, since they offer high energy densities, long runtime, instant recharging and use liquids as fuel. Ethanol emerges as a very attractive fuel when compared with the other alcohols commonly used in DFCs, since it is less toxic, has higher

[*] Corresponding Author's E-mail: vaniaso@fe.up.pt.
[†] Corresponding Author's E-mail: apinto@fe.up.pt.

energy density and can be produced from renewable sources. Based on that, direct ethanol fuel cells (DEFCs) emerge as a good choice for producing sustainable energy and have received great attention, in the last years, both in fundamental and technological aspects. However, they are still far from attaining acceptable levels of power output, since their performance is affected by the slow electrochemical ethanol oxidation and water and ethanol crossover.

This work provides a review on the experimental and modeling studies performed with DEFCs, towards its optimization. Therefore, this work starts with the fundamentals and main drawbacks of these systems, followed by an exhaustive review on the different studies performed regarding the effect of operating conditions (ethanol concentration, air pressure, anode and cathode flow rate and temperature) and design parameters (membrane electrode assembly characteristics/materials and flow field effect) on its power output and performance. As mathematical models are a fundamental tool for the design and optimization of fuel cells systems, this work also includes a section devoted to the modeling studies accomplished for DEFCs. Towards the use of DEFCs in portable electronic devices, different studies performed with passive direct ethanol fuel cells are presented at the end of this work.

Keywords: direct ethanol fuel cells, design considerations, modelling, operating conditions, passive, review

INTRODUCTION

Fuel cells have been identified as a promising power source for transportation and portable electronic devices, since they convert the chemical energy of a fuel directly into electrical energy, have low emissions and absence of moving parts [1]. The direct fuel cells (DFCs) use liquid fuels, have a compact design and potentially can offer up to ten times the energy density of rechargeable batteries. These advantages make the technology attractive to the rapid growing need for portable power sources. The disadvantages of using DFCs are: (i) the low anode reaction rates, resulting in higher activation losses and overpotentials and (ii) the fuel crossover towards the cathode, which decrease the fuel utilization, since some of the fuel feed to the anode do not react at this side, and the cathode performance and produces additional heat. Among all the investigated possible fuels, liquid alcohol fuels, which have high energy density and are easy to transport, store and handle become an alternative to hydrogen [2]. Up to a few years ago, methanol was the most used fuel and much progress has been made in the development and optimization of direct methanol fuel cells (DMFCs) [3-6]. However, methanol is toxic for human beings, is easily volatile and inflammable as well as nonrenewable [2]. Among the other alcohols available, ethanol appears as an attractive and promising fuel due to: (i) its non-toxicity, (ii) its natural availability, (iii) its renewability and (iv) its higher power density. Based on that, direct ethanol fuel cells (DEFCs) appear as a more attractive technology than DMFCs [2].

In DEFCs, some of the ethanol fed to the anode permeates through the membrane to the cathode, similar to the case of DMFCs. This leads to a decrease of the overall cell voltage due to the formation of an unwanted current at the cathode [7]. Compared to methanol, ethanol has low permeation rates mainly due to its larger molecular size, however work on this subject is fundamental to better understand the ethanol transport phenomena through the membrane in order to minimize them and achieve higher power outputs.

Another challenge of DEFCs is the cleavage of carbon-carbon bond, which is very difficult to break at low temperatures. Platinum (Pt) is the most active material, however if Pt is used alone self-inhibition occurs. Therefore, an additive is used to modify the surface of Pt. Among the different additives used, the more pronounced enhancement was observed with Tin (Sn) and Ruthenium (Ru) [8-23]. However, the DEFC performance is still poor and needs to be greatly improved.

In DEFCs, water is produced at the cathode, and to avoid some ethanol crossover problems, more dilute solutions are fed to the anode side. However, the presence of higher quantities of water on the cathode due to the oxygen reduction reaction along with water permeating from the anode due to electro-osmotic drag and diffusion tends to flood the cathode leading to an increase in oxygen transport resistance and reducing performance.

Table 1. Direct ethanol fuel cell reviews based on areas of investigation.

Feature	Reviews/year																					
	2006-2009				2010-2011				2012-2013					2014-2015			2016-2018					
	[7]	[8]	[9]	[10]	[11]	[12]	[13]	[14]	[15]	[16]	[17]	[18]	[19]	[20]	[21]	[22]	[23]	[24]	[25]	[26]	[27]	[28]
Anode catalyst	X	X	X	X	X	X	X	X	X	X	X	X	X	X	X	X	X			X		
Cathode catalyst	X	X	X	X	X	X	X	X	X	X	X	X	X	X	X	X				X		
Membrane	X				X	X	X		X							X			X	X	X	
Cell design and performance	X			X			X					X					X			X		
Ethanol crossover	X								X								X	X	X	X	X	X
AEM					X	X	X		X			X			X	X		X		X	X	
Operating conditions	X				X				X													
Non-Pt catalyst			X	X	X	X	X		X	X			X	X		X	X			X		
Alkaline media		X		X	X	X			X			X	X	X	X	X	X			X	X	
Heat management									X													X
Water management									X											X	X	
Durability and stability									X					X			X		X		X	
Modelling																		X		X		X
Applications									X					X	X		X					

Therefore, another important issue on DEFCs is how to eliminate water from the cathode and feed water to the anode to compensate water loss due to crossover through the membrane. Although, some efforts have been made for studying water transport in DMFCs [3, 4], a general understanding of water crossover in DEFCs is far less understood. Based on these challenges, there has been a growing amount of work in the last years, both on the operating conditions (ethanol and oxygen concentration, temperature and flow rates) and on the design ones (catalyst materials and loadings, diffusion layers materials), which managed to increase power outputs [7-28].

In the last years, advances have been made in the DEFC research and some review articles on this type of fuel cell are available, each one with a different accent or emphasis (Table 1). These include the following: anode and cathode catalysts [8-23, 26], membrane electrode assembly (MEA) preparation procedure and materials [7, 11-13, 15, 22, 25-27], cell design and performance [7, 10, 13, 18, 23, 26], ethanol crossover [7, 15, 22, 23, 25-28], AEM-DEFCs (anion exchange membrane direct ethanol fuel cells) [11-13, 15, 18, 21, 22, 24, 26, 27], operating conditions [7, 11, 15], non-platinum catalysts [9-13, 15, 16, 19, 20, 22, 23, 26], alkaline media [8, 10-12, 15, 18-23, 26, 27], heat management [15, 28], water management [15, 26, 27], durability and stability [15, 20, 23, 25, 27], applications [15, 20, 21, 23] and modelling [24, 26, 28].

Based on the review articles already published and described above and due to a lack of recent work regarding experimental and modelling studies on DEFCs, the main goal of this paper is to provide a recent review concerning these issues. We hope that this review will be highly useful for readers who are interested in beginning their work in this area, as well as, to the researches already working in this field to access the progress achieved until now.

OPERATING CONDITIONS

Operating conditions like reactants flow rates and concentrations and fuel cell temperature are key parameters for finding optimal performances.

An understanding of the interdependence of the operating conditions emerges as an important role on DEFCs optimization. In the last years, some work had be done in order to achieve optimal performances with higher power outputs [29-43].

Ethanol Concentration

The influence of ethanol concentration on the DEFC performance generally reflects different phenomena. An increase on ethanol concentration leads to an increase on the coverage of the electrocatalyst sites, presented on the anode catalyst layer, and to an increase on the limiting current density. Since, the polarization behavior in the mass transfer region is directly related to the ethanol concentration, an increase in the limiting current density is achieved with an increase in ethanol concentration. Almost all the studies reported in literature, regarding the effect of ethanol concentration on fuel cell performance, showed that there is a general increase in the limiting current with an increase in concentration whilst at low current densities higher ethanol concentrations lead to a reduction in the cell voltage [29-34, 36-38].

It should also be mentioned, that in DEFCs, such as in DMFCs, some of the fuel feed to the anode crosses the membrane to the cathode due to the inability of the membranes to prevent the fuel from permeating its polymer structure [39]. Diffusion, due to the fuel concentration gradient among both sides, and electro-osmotic drag are the prime driving forces for fuel transport through the membrane. The fuel that reaches the cathode side is attached to the cathode catalyst decreasing the catalytic sites available for oxygen reduction reactions and forming a mixed potential at this side, which decreases the fuel cell efficiency and lowers the energy density of the system. In addition, the exothermic reaction that occurs between the ethanol and the oxygen on the cathode generates extra heat. Therefore, to enhance the fuel cell performance it is necessary to reduce or even eliminate the loss of fuel across the membrane. Ethanol should have lower permeation rates than methanol due to its larger molecular size, however work on this subject found that the DEFCs also have lost in performances due to ethanol crossover [29, 30, 32, 37, 39-41]. In DEFCs, the open-circuit voltage is much lower than the ideal voltage due to ethanol crossover and it is lower for higher ethanol feed concentrations [34]. This is due to the fact an increase on ethanol concentration leads to a higher concentration gradient between both sides with a consequent increase in the ethanol flux through the membrane [37]. Hence, a higher ethanol concentration originates higher mixed potentials at the cathode side and lower fuel cell power outputs.

On the anode side of a DEFC, carbon dioxide is produced as a result of the ethanol oxidation reaction. If the carbon dioxide cannot be removed efficiently from the surface of the catalyst and diffusion layers it remains covering these surfaces and consequently induce a decrease of the effective mass transfer area. Therefore, gas management on the anode side is an important and critical issue in DEFCs [34, 42]. Katayanagi and Yamazaki [34] present an experimental study on the effect of fuel cell operating conditions on carbon dioxide generation. They found that a decrease on ethanol concentration in the fuel leads to an increase on carbon dioxide rate. They demonstrated that more carbon dioxide was produced with a higher concentration polarization. The results indicated that the production of carbon dioxide is promoted as the concentration of ethanol or intermediate products decreases on the anode surface.

As can be seen, select the best ethanol concentration for a DEFC requires a delicate balance among the effects on ethanol oxidation kinetics, on ethanol crossover and on carbon dioxide production to enhance the cell performance. It is reported in the literature that the highest performance is achieved with an ethanol concentration of 2M, since for higher ethanol concentrations the negative effect on ethanol crossover is higher than the positive ones.

Air Pressure

In the experimental studies performed with DEFCs high pressure operation was chosen in order to achieve a high performance. Higher pressures increase the oxygen concentration in the cathode side, so more oxygen reacts at the cathode catalyst layer, enhancing the oxygen reduction reaction and consequently reducing the activation overpotential [35, 36]. It was also found that the crossover of ethanol can be, slightly reduced by a high cathode air pressure, reducing the poisoning of the cathode active sites by ethanol and enhancing the oxygen reduction reaction. Additionally, increasing the oxygen concentration, by increasing the air pressure, in the cathode gas significantly increases the carbon dioxide generation ratio on the anode side, due to an increase of the oxygen crossover from the cathode to the anode side. The oxygen that reaches the anode via the electrolyte membrane promotes the carbon dioxide production reaction [34]. Also low pressures are needed for portable applications, one of the most important segment markets for this type of fuel cells. Therefore, it is crucial to find optimal performances with low pressure values. Some work with other alcohol fuel cells suggested a high performance even at atmospheric pressure by optimizing the structures and compositions of the different fuel cell layers [3].

Anode and Cathode Flow Rate

Another important aspect of the DEFC is the possibility of water flooding at the cathode pores and channel structure due to water crossover through the membrane and to the water production by the cathode reaction. The water production within the cathode catalyst layer and its transport through the cathode diffusion layer add another mass transport resistance for oxygen on its flow towards the reaction zone, and therefore reduces the limiting cell current with respect to the cathodic reaction. Therefore, an efficient removal of carbon dioxide gas bubbles and liquid water produced must be maintained to allow the access to catalyst sites. Removal of carbon dioxide bubbles and prevention of cathode flooding can be achieved by increasing flow rates [30, 34, 35]. When a DEFC is operating under low flow rates the oxygen and the ethanol concentration decrease along the flow channels resulting in an insufficient mass transport of fuel and oxidant to the reaction zone. This will lead to an insufficient fuel and oxidant supply and the fuel cell will be faced with lack of reactants. Higher flow rates will increase the mass flux to the diffusion and catalyst layers increasing the anode and cathode reactions and the fuel cell performance. However, increasing flow rates requires more pumping power and a very high flow rate at the cathode will dry out the polymer membrane, decreasing proton conductivity and hence cell performance.

Different flow rates were proposed for DEFCs by the available literature. Generally, fuel cell performance increases with an increase of ethanol and air flow rate but just to a certain optimal value, after which the fuel cell performance is not affected by the reactants flow rate [30, 35, 36].

Heat Management

Controlling the operating temperature and the heat transfer management are critical to obtain optimized performances. Higher temperatures lead to an increase in cell performance, since the electrochemical kinetics at the anode and cathode are favored by increasing temperature. Therefore, most of the studies with DEFCs were performed at high temperatures [29-33, 35, 36]. However, higher cell temperatures also have negative effects: the rate of ethanol crossover and the water loss from the anode to the cathode increase with temperature and the membrane stability and the oxygen partial pressure decrease with temperature [29, 32]. The effective influence of the cell temperature is, in this way, a result of both positive effects on the electrode kinetics and the combined negative effects. The increase of temperature also increases the open-circuit voltage and reduces the activation overvoltage according to the Arrhenius relation, thus resulting in a higher performance [35]. However, if the operation temperature is similar to the boiling temperature of the solution, the cell performance decreases, since small bubbles of the vapor formed in the catalyst and diffusion layers may obstruct the fuel transport [31].

Commonly, the fuel cell performance is improved with higher temperatures. However, for portable applications it is advantageous to work at room temperature, so high temperatures are not allowed. The main research target is to operate a DEFC at ambient temperatures with optimized performances [43].

Design Considerations

To increase the performance of a DEFC it is extremely important to reduce the kinetic, ohmic, mass and heat transport limitations and reduce the transport of the different species through the membrane. Based on that, some work was performed in order to improve the membrane, the catalyst and the diffusion layers characteristics and properties [29-32, 35, 39, 44-61] and to found an optimized design for the reactants flow field [62, 63].

Membrane Electrode Assembly

The MEA composed by the catalytic layers, diffusion layers and the proton exchange membrane (PEM) is the most important part of a fuel cell.

Membranes

The PEM must have relatively high proton conductivity, must be a barrier to the mixing of fuel and reactant and must be chemically and mechanically stable in the fuel cell environment. Typically, the membranes for fuel cells are made of perfluorocarbon-sulfonic acid ionomer, which results from the combination of tetrafluorethylene with perfluorosulfonate monomers, commonly known as Nafion membranes. The sulfonic acid groups, presented in the Nafion chemical structure, are ionically bonded and are responsible for creating hydrophilic regions, thus absorbing relatively large amounts of water. The proton movement within the well-hydrated regions makes these materials proton conductive. Nafion membranes are available in several thicknesses and are the most commonly used in DEFCs. However, as already mentioned, Nafion membranes were found to have high ethanol permeability, allowing the transport of ethanol molecules through its structure to the cathode side and oxygen crossover towards the anode side [44]. These parasitic processes decrease fuel cell performance due to their reaction on the wrong side of the fuel cell with a consequent loss of fuel and oxidant. Since one of the major goals on fuel cell developments and optimization is to reduce fuel and oxidant loss, some solutions have been proposed, which include using different membrane thicknesses or using membranes with different compositions [32, 45-53]. Regarding the membrane thickness, Andreadis et al. [32] found that, for low current densities, the thicker membrane (Nafion 117) leads to higher performances while, for high current densities, the better performance was achieve with the thinner membrane (Nafion 112). This is due to the fact that at low current densities the fuel cell suffers from large amounts of ethanol crossover and thicker membranes have low ethanol crossover rates due to higher mass transfer resistances imposed by the membrane. In such conditions, ethanol has to overcome a higher pathway to reach the cathode side, so thicker membranes have positive effects on its performances. At higher current, where the fuel crossover rate is not the most significant loss in performance, thinner membranes lead to higher performance since conduct to small internal cell resistances. Other attempts to modify or adopt new electrolyte materials to avoid or at least decrease ethanol crossover have been reported in the last years [45-53]. Fu et al. [45] developed a series of hybrid proton-conducting membranes with an interpenetrating polymer network (IPN) structure that exhibited high proton conductivity, less swelling and lower ethanol permeability than Nafion membranes. However, since the membranes are hydrophilic they could have an increased water flow from the anode to the cathode, requiring a close look of this effect on water management under fuel cell operation. Barbora et al. [46] prepared Nafion/Titanium dioxide composite membranes, by casting method, with different titanium dioxide loadings and found that these composite membranes show a decrease in ethanol crossover of 71% for 5% of titanium dioxide and the proton conductivity was highest for 1% of titanium dioxide. They concluded that the composite membrane prepared with 5% of titanium dioxide has the maximum characteristic factor for ethanol and thus can be suitable for DEFC applications. The sulfonated poly(ether ether ketone)

(SPEEK) membranes have been recently studied in order to evaluate their proton conductivity and fuel permeability as potential application in DEFCs [47-51]. The advantages of using SPEEK membranes over the Nafion are their excellent chemical resistance, high thermo-oxidative stability and low cost. SPEEK membranes are typically obtained by either directly introducing sulfonic acid groups onto the polymer backbone or directly polymerizing sulfonated monomers. For SPEEK membranes it has been reported that they have a higher water/ethanol selectivity and lower water and ethanol permeability than Nafion membranes [47-51]. Proton-conduction membranes composed of poly(vinyl alcohol) (PVA), phosphotungstic acid (HPW), (diethylenetriamine)pentaacetic acid (DTPA) and poly vinylidene fluoride-hexa fluoro propylene (PVdF-HFP) have been prepared and tested as a possible solution to mitigate the ethanol crossover effect [52-54]. Phosphotungstic acid has been described as a good proton conductor and has high thermal stability, however has high solubility in water, leading to HPW loss together with water at low temperatures [52, 54]. Based on that, Tang et al. [52] developed a HPW-silica mesoporous nanocomposites membrane, which showed high proton conductivity, good performance, low ethanol permeability and excellent water retention capacity. PVA is other interesting polymer since it was found to have good chemical resistance, good mechanical properties, reduced ethanol permeability through the membrane and low cost. However, the absence of specific functional groups responsible for conducting protons in this polymer, limits the use of PVA alone therefore, needs to be tailored with other materials that promote proton conductivity. Among the different modifications available for PVA, Gomes et al. [54] developed a hybrid membrane with acid groups such as HPW into PVA matrix and crosslink with DTPA, which has a high affinity for metal cations, and is a good chelating agent. The resulting membrane presented an ethanol permeability two orders of magnitude smaller than with Nafion membranes and with a high proton conductivity.

Catalyst Layers

As already mentioned, on either side of the membrane are placed the anode and cathode catalyst layers. In the reactions occurring in a DEFC, gaseous and liquid reactants, electrons and protons are presented. The reactions take place in a portion of the catalyst surface where all three species have access. Electrons travel through electrically conductive solids, protons through ionomer and reactants through voids. Therefore, the catalyst layers must be porous to allow the reactants access to the reaction sites and to remove the products formed in the electrochemical reactions. The most commonly used catalyst for a fuel cell is Platinum (Pt) based catalyst. However, platinum is very expensive which limit its application in fuel cells since they are intended to be market competitive. Additionally, ethanol is an alcohol with a carbon-carbon (C-C) bond on its structure and to achieve a complete oxidation of ethanol to carbon dioxide, producing twelve electrons per molecule, C-C cleavage must occur. Incomplete C-C bond cleavage results in partial oxidation pathways and the formation of acetaldehyde and acetic acid [55]. These species desorb from the surface and diffuse to the bulk electrolyte, therefore, do not undergo further oxidation, resulting in a loss of efficiency since only a small fraction of ethanol is fully oxidized. Previous work shows that ethanol cannot be totally oxidized at temperatures below 100°C. Working at low temperatures is essential for real applications, therefore the ethanol oxidation represents a great challenge in DEFC research. Different experimental studies and review papers can be found in literature regarding this issue and different catalyst have been proposed, however, until now, only a few provide complete oxidation [7-27, 31]. The catalysts generally used are Pt-Ru (Platinum/Ruthenium) or Pt-Sn (Platinum/Tin) at the anode side and Pt at the cathode side with loading from 1 to 4 mg/cm2 at the anode and cathode side. The structure and thickness of the catalyst layer were also found very important for the kinetics of the reactions and for the diffusion of the different species [29, 31, 32, 56, 57]. Andreadis et al. [29, 32] evaluated the influence of the catalyst layer thickness on ethanol transport through the membrane, for different fuel concentrations. The results showed that an increase on the catalyst layer thickness results in a decrease of the ethanol flux towards the cathode, since thicker catalyst layers impose higher mass transfer resistances to ethanol and more ethanol can be involved in the anodic reaction. Some studies concerning the effect of the catalyst loading on the DEFC power output showed that higher loadings lead to higher specific surface areas, resulting in higher performances [31, 32, 56]. However, it was found that an increase on the anode catalyst loading has a larger impact on the power density than an equal one on the cathode side. This can be explained by the fact that the ethanol oxidation reaction is much slower than the oxygen reduction reaction. So, a higher anode reaction area has a more substantial impact on the overall cell performance [32]. Based on these results and with the goal of reducing the catalyst loading to achieve lower cost without sacrifice the fuel cell performance, Basu et al. [58] developed a DEFC with a lower anode and cathode catalyst loading using sulfuric acid in ethanol and a Ni-mesh as current collector at the anode. The results showed and enhance on the fuel cell performance by 40-50%, since the use of sulfuric acid in ethanol and the Ni-mesh as current collector offsets the decrease in performance at low catalyst loadings by increasing the proton conductivity through the electrode and membrane and improving the catalytic activity of the anode. Wang et al. [59] presented double-layered anode catalyst layers to improve the performance of a DEFC. The results showed that the double-layered anode catalyst layer had higher product yields, current efficiencies, ethanol oxidation efficiencies and DEFC performance than a single anode catalyst layer. Wan et al. [60] developed a composite anode comprising an inner catalyst layer containing a thin layer of Pt-Sn nanoparticles directly deposited on the Nafion membrane. The results showed that the proposed anode reduce the ethanol crossover up to 17% and yield a 6% better performance than a normal MEA. It was also found that the inner catalyst layer serves to suppress the ethanol crossover and to improve the ethanol oxidation reaction and the fuel cell performance. As already mentioned, another parameter characteristic of the catalyst layers that has a clearly effect of the fuel cell performance is the porosity, since the catalyst layers should be sufficiently porous to allow fuel supply to the active sites and efficient product removal. Andreadis et al. [32] found that with an increase on the catalyst layer porosity more reactant reach the active sites and participate in the electrochemical reactions but more ethanol reaches the membrane with a consequent increase on the fuel crossover rate. However, the results showed that the benefit of an increase on catalyst porosity on the electrochemical reactions is higher than the negative effect on crossover. Trimetallic catalysts have also been proposed by some authors since it was found that they exhibit an enhanced electrochemical activity than the bimetallic ones [22, 57]. Beyhan et al. [57] found that Pt-Sn-Co and Pt-Sn-Ni catalysts have a higher activity for the ethanol oxidation reaction than Pt-Sn, demonstrated by their higher performances and maximum power densities.

Diffusion Layers

The layers between the catalyst layer and the flow field plates are called the diffusion layers. Although, these layers do not participate directly in the electrochemical reactions they allow the access of reactants to the catalyst active area, provide the products removal from the catalyst layer, electrically connect the catalyst layer to the flow field plate, allow heat removal and provide mechanical support to the MEA, preventing it from sagging into the flow field channels. Based on that, the diffusion layers must be porous and electrically and thermically conductive. These requirements are best met with carbon fiber based materials such as carbon fiber papers and woven carbon fabrics or cloths [35, 61]. Different properties of the diffusion layers affect the fuel cell performance such as the layer thicknesses, the porosity, which influences the species transport, the wettability and the roughness. Different diffusion media with different structural parameters result in different transport characteristics [3, 4]. Andreadis et al. [32] studied the effect of the porosity and thickness of the diffusion layers on fuel cell performance and on the formation of the undesirable current on the cathode side. Regarding the effect of the porosity it was found that an increase on porosity leads to an increase of ethanol and oxygen concentration in the catalyst layers with a consequent increase on the electrochemical reaction rates and a decrease on the anode and cathode overpotentials. However, a higher ethanol concentration on the anode catalyst layer leads to a higher concentration gradient between both sides with a consequent increase on the ethanol flux through the membrane [32]. Concerning the effect of the diffusion layer thickness it was found that an increase on thickness leads to a decrease on fuel cell performance due to a decrease of ethanol and oxygen concentration on the catalyst layer surface which increases the activation overpotentials [32]. Nevertheless, thicker diffusion layers have low ethanol crossover rates.

The most commonly used materials as diffusion layers in fuel cells are carbon cloth and carbon paper [3, 4, 35, 61]. Carbon cloth is more porous, less tortuous and thicker than carbon paper, although carbon paper has excellent electronic conductivity [3, 4, 35]. It should, here, be mentioned that on the anode side gaseous carbon dioxide is produced by the anode reaction and liquid water is produced in the cathode reaction, which must be efficiently removed from the catalyst layers. If the carbon dioxide and the water cannot be removed from the catalyst surface they cover the surface decreasing the effective active area for the electrochemical reactions. The texture or carbon paper with a highly tortuous structure enhances the interactions between the water and carbon dioxide and the solid. Hence, these species tend to remain attached to the surface leading to a blockage of the active sites lowering the ethanol oxidation and oxygen reduction reaction rates and the cell performance. In addition, carbon paper with its high tortuosity imposes a higher oxygen and ethanol transport resistance with lower ethanol and oxygen concentrations in the surface of the catalyst layers [3, 4, 35]. In order to solve the water attachment problems the diffusion layers are PTFE (polytetrafluoroethylene) treated making these layers hydrophobic. Although, the anode diffusion layer should be hydrophilic to facilitate the mass transfer of the dilute ethanol solution to the anode. Studies regarding this issue show that the use of a diffusion layer hydrophobic on the anode side leads to higher cell performances, since the PTFE treatment created a higher region for gas movement, reducing the liquid water accumulation and leading to higher cell performances [61].

Another way to manage the mass transport problems in DEFCs, is to use an anode microporous layer (MPL) between the catalyst layer and the diffusion layer [32, 61]. It is expected that the presence of this additional layer can reduce the ethanol crossover rate and decrease the two-phase transport due to a greater micro-pore distribution of the MPL. However, it was found that the existence of a MPL layer on the anode side increases the mass transfer resistance, decreasing the ethanol concentration on the catalyst surface. This leads to a decrease on the ethanol oxidation rate with a consequent decrease on the fuel cell performance despite the fact that a decrease on the ethanol crossover rate was achieved [32, 61]. The results suggest that the kinetic influence is more important than the ethanol crossover effect.

Flow Field Effect

The performance of a DEFC depends not only on the types of materials used but also on the design of the flow plates, since it is very important to distribute uniformly the reactants all over the cell active area, as well as, to remove the products formed on the electrochemical reactions that take place on both cell sides. This is performed by the flow plates, which have flow fields with different designs, such as channels all over the plate, channels in some patterns or porous structures. As can be seen in Figure 1 different types of flow pattern designs are available for fuel cells, such as parallel, spot, serpentine and interdigitated channels [62, 63].

The parallel and the spot designs usually give comparable performances, although the use of a spot design has some practical limitations, such as, its fabrication, which is not easy and consequently has higher costs. These two designs have low pressure drops, but have an inhomogeneous reactant distribution and some difficulties in removing the reaction products. In the serpentine design the reactant is forced to flow in a zigzag way along parallel channels and the reactant molecules have access to the catalytic sites through diffusion across the diffusion layers. This design is helpful to eliminate the products formed on the anode and cathode reactions and to improve species transport, however suffers from high pressure drops all over the channels. In the interdigitated design the reactant is forced to enter into the electrode pores and exit from them under a pressure gradient. In this way the flow is no more governed by a diffusion mechanism but becomes a forced convection mechanism. The mass transport is enhanced but a high pressure difference is required between channels.

In general, different designs have positive and negative effects on the fuel cell operation and consequently on its performance. A proper selection of the flow field with respect to the DEFCs operational conditions can help to achieve cost and performance goals [62, 63].

MODELLING STUDIES

The DEFC is a multifaceted technology comprising electrochemical reactions, charge, mass and energy transfer phenomena. Since several simultaneous processes occur in DEFC, the necessary formulation of adequate mathematical models is based on the need to reach a deeper understanding of internal processes (reactions, mass transport, heat transport) which cannot be observed directly in experiments or the possibility to apply mathematical optimization methods to obtain hints as to where further technical improvements may be required or beneficial, and, finally, for the development of optimal control and operating strategies. Therefore, the development of mathematical models is critical to the design and optimization of fuel cells. Based on that, fuel cell modelling has received much attention over the last years and different types of approaches are available in the literature [64]. Nowadays, with the available computational power, the model simulations are getting more complex, including a high level of details directly related to the basic fuel cell modeling features and to the simplifications and assumptions that are usually made by the authors. It is important to understand all the assumptions made, in order to recognize the limitations and to accurately interpret the results provided by each model. The majority of the fuel cell models aims to describe the interactions occurring among the several physical and electrochemical phenomena within the different layers of the cell, and some studies were performed for DEFCs [29, 33, 37, 40, 65-69].

Andreadis et al. [29] reported a simple one-dimensional (1D), steady-state, isothermal and single phase mathematical model to describe the mass transport phenomena and the performance of a DEFC. The ethanol transport in the diffusion layer was assumed as a combination of diffusion and convection, while in the catalyst layer the ethanol oxidation was described by the Tafel equation. For the sake of simplification, the water transport through the membrane was based on the assumption that both anode and cathode are fully hydrated. The model was used to predict the effect of operating parameters on the cell performance, and it was found a relatively good agreement between theory and experimental results related to both ethanol crossover rates and DEFC performance. Years later, Andreadis et al. [65] developed a 1D, steady-state and isothermal model based on the previous work [29] but included the equations related to the modeling of the humidified oxygen transport and its reduction on the cathode side, the formulation for the concentration polarization at both anode and cathode, the direct correlation between ethanol crossover and the unwanted current at the cathode side, the effect of the mixed potential and the oxygen feed concentration on the DEFC performance. According to the model results, when the unwanted current is considered in the model development the most evident effects are on the reduction of the open circuit voltage and on the discharge behavior.

Pramanik et al. [33] worked on a mathematical model for DEFCs considering the effect of the distinct overpotentials. In this model, the activation overpotential term is formulated considering the Butler–Volmer equation, and the ohmic overpotential is modeled based on the proton conductivity of the membrane and the ohmic losses on the electrodes and current collectors and on its interface. Furthermore, the concentration overpotential is formulated using Fick's law, the modified Butler–Volmer equation and transport process through the electrodes and electro-catalyst layers. Using this model, the authors were able to predict the experimental data on current–voltage with reasonable agreement considering temperatures in the range of 40ºC to 60°C.

Suresh et al. [37] proposed a comprehensive 1D, single phase and isothermal model, which accounts the mass transport and the electrochemical reactions on both DEFC sides. Tafel equation was used to describe the ethanol oxidation reaction on the anode and cathode side due to the ethanol crossover and the oxygen reduction reaction. The model fully accounts for the formation of an unwanted current on the cathode due to ethanol crossover. The model was validated using very limited data from the literature, it predicts reasonably well the limiting current and the polarization behavior for different cell parameters, as well as, the effect of key operating conditions on the ethanol transport through the membrane and on the cathode overpotential.

Meyer et al. [40] developed a 1D model where the ethanol oxidation reaction in the catalyst layers is described by different reaction steps. The authors assume a bifunctional reaction mechanism for the water activation on a binary catalyst, and the reactions are combined with the charge and mass transport. Additionally, the model considers the reactants crossover through the membrane, and the formation of mixed potentials on the anode and cathode side due to the undesirable oxidation of ethanol and oxygen reduction, respectively. Although this is a very interesting approach, the model still needs to be validated with further experimental data.

Specific modeling and simulation of the DEFC anode were also investigated as an attempt to better understand the anodic process [66, 67]. Andreadis et al. [66] developed a mathematical model that explains, in one phase and one dimension, the mass transport of ethanol throughout the anode compartment and proton exchange membrane. The cell operates under steady-state conditions, and the driving force for both water and ethanol transport through the catalyst layer consists on the diffusion and the electro-osmotic drag. The effect of current density, temperature and ethanol concentration on both ethanol crossover and anode overpotential were investigated, and the authors found a good agreement between the simulated and experimental results. Sousa Jr. et al. [67] proposed a mechanistic model that could supply details of the physical system, such as the concentration profiles of the various intermediates generated in the anode and the coverage of the adsorbed species on the electrode surface, as well as, the anode overpotential–current relationship. This model uses the Navier-Stokes equations for fluid flow and the Darcy's Law for porous media flow, and the heat transfer is modeled considering diffusion, convection and conduction. For the cathode, a standard model is used. The authors found similar results and trends for the model simulations and the experimental results. In their work, Goel and Basu [68] developed a one-dimensional steady state and isothermal model to study the electrochemical behavior of a DEFC. This model takes into account the mass transport relations for the different species transport in the flow channels, diffusion and catalyst layers, the multistep ethanol oxidation reactions at the anode catalyst and the electrochemical reduction reaction at the cathode side and proton, water and ethanol transport through the membrane. The model was used to predict the cell performance for different anode catalyst, Pt-Re-Sn/C, Pt-Ru/MCN, Pt-Re-Sn/MCN and Pt-Re-Sn/t-MWCNT catalysts and different temperatures (80°C, 90°C and 100°C). The model was also used to predict different polarization losses and dependence of ethanol crossover on ethanol concentration and current density. Recently, Souza et al. [69] presented a two-dimensional (2D) model, which considers the flow and species concentration as a function of time and space (fuel consumption and products formation) on the anode and cathode sides of a DEFC. The model also considers the overpotential losses and the ethanol transport through the membrane. They conclude that the simulation results regarding the effects of the different operating temperatures, catalysts and ethanol concentrations on the DEFC performance are in agreement with the experimental data presented in literature.

PASSIVE DEFCs

The DEFCs can be categorized as active and passive, according to the method used to provide the reactants to the cell [30, 70]. Active systems use external components to feed the fuel and oxidant to the cell, such as pumps and blowers/compressors, requiring power to operate. An active feeding improves the mass transport and the electrochemical activity of the cell, enabling higher fuel cell performances. However, the use of external power and moving parts leads to a higher cost and complex system making difficult its implementation in portable devices. Passive systems emerge as promising candidates for portable applications due to a simpler and compact design since the transport of reactants and products occurs by natural capillary forces, diffusion, convection and evaporation [70]. This system does not require moving components and power to operate, since the ethanol solution is stored in a reservoir presented on the anode side and the cathode side of the cell is exposed to the atmosphere. However, passive systems have lower power outputs, since their performance is affected by the slow electrochemical ethanol oxidation reaction and water accumulation or flooding at the cathode, which leads to a decrease of the cathode reduction reaction rate [30, 70-72]. Nevertheless, the cell performance can be improved by using different ethanol concentrations and different materials for the membrane, diffusion and catalysts layers [70-72]. Pereira et al. [71] studied the effect of different operating and design parameters on the performance of a passive DEFC. The maximum power output was achieved when the cell was operated with a thicker membrane, a higher anode catalyst loading, ELAT as anode diffusion layer and carbon cloth as cathode diffusion layer and an ethanol concentration of 2 M.

As already mentioned the development of mathematical models is crucial to overcome the different challenges of passive DEFCs and studying the impact of the different operating and design parameters on the fuel cell performance. Oliveira et al. [72] developed a steady state and one-dimensional model considering the electrochemical reactions that occur on both sides of a passive DEFC and the heat and mass transport phenomena. The model demonstrated a good agreement between theory and experimental results, especially for low current densities. For high current densities, the model predictions overlap the experimental behavior of the cell due to the incapability to predict the effect of the two-phase flow on both anode and cathode sides, due to the formation of gaseous carbon dioxide on the anode and water drops at the cathode. Nevertheless, this mathematical model can be a useful tool to optimize the performance of a passive DEFC.

Conclusion

In the last years, some efforts have been made in the development of DEFCs to enhance their performance. Some studies have focused on analysis and improvement of single parts of DEFC systems and different materials, designs and operating conditions have been suggested. Although the results of these studies are useful, it is important to note that materials and conditions optimized for one type of fuel cell may not be always recommended for another. In this way, when developing a DEFC, the materials and conditions need to be carefully chosen in order to achieve optimal performances. Regarding the operation conditions, it is possible to enhance performance using different ethanol concentrations, flow rates and temperatures. A great challenge on the development of this type of fuel cell, in order to be used in real applications, is the development of cost-effective materials. To achieve that, different materials must be explored as current collectors, membrane, and diffusion and catalyst layers. Additional studies on cost-efficient electrode materials and optimization of configurations are needed.

One of the great obstacles in DEFC developments is the ethanol crossover, since ethanol diffuses through the membrane generating heat but no power. This can be limited if the cell operates with low ethanol concentration on the anode. However, this significantly reduces the system energy density since large amounts of water are presented in the system and water will produce no power. In addition creates higher water concentration gradients, so more water crosses the membrane. The existence of large amounts of water on the cathode, due to water crossover and production on the cathode reactions, may lead to cathode flooding, with a consequent decrease on DEFC performance. The solution is to use high concentrations of ethanol and use different membrane, diffusion layers and catalyst layers materials with different thicknesses.

Mathematical models appear as a useful tool to understand and predict the main transport phenomena and electrochemical processes helping in fuel cells optimization in terms of design and power output. Despite some work done in DEFC modelling, there is still the need for reliable mathematical models of complete fuel cells coupling, simultaneously, ethanol, water and heat transport processes. Special attention should be devoted to the inclusion of ethanol crossover because it affects negatively the fuel cell performance.

Such models are helpful for the discovery of new cell designs and operation regimes of the DEFC systems.

Passive DEFCs are very promising systems to be implemented in portable electronic devices, since are less expensive, simpler and compact than active ones. However, more studies are needed to clearly access the influence of the different materials, designs and operating conditions on the cell power output towards its enhancement.

ACKNOWLEDGMENTS

This research is being supported by the Project "HYLANTIC"–EAPA_204/2016, which is co-financed by the European Regional Development Found in the framework of the Interreg Atlantic Programme and the Project PTDC/NewPortCell-POCI-01-0145-FEDER-032116 - funded by FEDER funds through COMPETE2020 - Programa Operacional Competitividade e Internacionalização (POCI) and by national funds (PIDDAC) through FCT/MCTES. POCI (FEDER) also supported this work via CEFT.

REFERENCES

[1] Sundmacher, K. (2010). Fuel cell engineering: toward the design of efficient electrochemical power plants. *Industrial & Engineering Chemistry Research,* 49 (21): 10159–82. https://doi.org/10.1021/ie100902t.

[2] Demirci, U. B. (2007). Direct liquid-feed fuel cells: thermodynamic and environmental concerns. *Journal of Power Sources,* 169 (2): 239–46. https://doi.org/10.1016/J.JPOWSOUR.2007.03.050.

[3] Oliveira, V. B., Rangel, C. M. and Pinto, A. M. F. R. (2009). Water management in direct methanol fuel cells. *International Journal of Hydrogen Energy,* 34 (19): 8245–56. https://doi.org/10.1016/J.IJHYDENE.2009.07.111.

[4] Oliveira, V. B., Falcão, D. S., Rangel, C. M. and Pinto, A. M. F. R. (2013). Water management in a passive direct methanol fuel cell. *International Journal of Energy Research,* 37 (9): 991–1001. https://doi.org/10.1002/er.2902.

[5] Falcão, D. S., Oliveira, V. B., Rangel, C. M. and Pinto, A. M. F. R. (2014). Review on micro-direct methanol fuel cells. *Renewable and Sustainable Energy Reviews,* 34: 58–70. https://doi.org/10.1016/J.RSER.2014.03.004.

[6] Awang, N., Ismail, A. F., Jaafar, J., Matsuura, T., Junoh, H., Othman, M. H. D. and Rahman, M. A. (2015). Functionalization of polymeric materials as a high performance membrane for direct methanol fuel cell: A review. *Reactive and Functional Polymers,* 86: 248–58. https://doi.org/10.1016/J.REACTFUNCTPOLYM.2014.09.019.

[7] Song, S. and Tsiakaras, P. (2006). Recent progress in direct ethanol proton exchange membrane fuel cells (DE-PEMFCs). *Applied Catalysis B: Environmental,* 63 (3–4): 187–93. https://doi.org/10.1016/J.APCATB.2005.09.018.

[8] Antolini, E. (2007). Catalysts for direct ethanol fuel cells. *Journal of Power Sources,* 170 (1): 1–12. https://doi.org/10.1016/J.JPOWSOUR.2007.04.009.

[9] Demirci, U. B. (2007). Theoretical means for searching bimetallic alloys as anode electrocatalysts for direct liquid-feed fuel cells. *Journal of Power Sources,* 173 (1): 11–18. https://doi.org/10.1016/J.JPOWSOUR.2007.04.069.

[10] Bianchini, C. and Shen, P. K. (2009). Palladium-Based Electrocatalysts for Alcohol Oxidation in Half Cells and in Direct Alcohol Fuel Cells. *Chemical Reviews,* 109 (9): 4183–206. https://doi.org/10.1021/cr9000995.

[11] Yu, E. H., Krewer, U. and Scott, K. (2010). Principles and Materials Aspects of Direct Alkaline Alcohol Fuel Cells. *Energies,* 3 (8): 1499–528. https://doi.org/10.3390/en3081499.

[12] Antolini, E. and Gonzalez, E. R. (2010). Alkaline direct alcohol fuel cells. *Journal of Power Sources,* 195 (11): 3431–50. https://doi.org/10.1016/J.JPOWSOUR.2009.11.145.

[13] Zhao, T. S., Li, Y. S. and Shen, S. Y. (2010). Anion-exchange membrane direct ethanol fuel cells: Status and perspective. *Frontiers of Energy and Power Engineering in China,* 4 (4): 443–58. https://doi.org/10.1007/s11708-010-0127-5.

[14] Antolini, E. and Gonzalez, E. R. (2011). Effect of synthesis method and structural characteristics of Pt–Sn fuel cell catalysts on the electro-oxidation of CH_3OH and CH_3CH_2OH in acid medium. *Catalysis Today,* 160 (1): 28–38. https://doi.org/10.1016/J.CATTOD.2010.07.018.

[15] Kamarudin, M. Z. F., Kamarudin, S. K., Masdar, M. S. and Daud, W. R. W. (2013). Review: Direct ethanol fuel cells. *International Journal of Hydrogen Energy,* 38 (22): 9438–53. https://doi.org/10.1016/J.IJHYDENE.2012.07.059.

[16] Brouzgou, A., Song, S. Q. and Tsiakaras, P. (2012). Low and non-platinum electrocatalysts for PEMFCs: Current status, challenges and prospects. *Applied Catalysis B: Environmental,* 127: 371–88. https://doi.org/10.1016/J.APCATB.2012.08.031.

[17] Braunchweig, B., Hibbitts, D., Neurock, M. and Wieckowski, A. (2013). Electrocatalysis: A direct alcohol fuel cell and surface science perspective. *Catalysis Today,* 202: 197–209. https://doi.org/10.1016/J.CATTOD.2012.08.013.

[18] Brouzgou, A., Podias, A. and Tsiakaras, P. (2013). PEMFCs and AEMFCs directly fed with ethanol: a current status comparative review. *Journal of Applied Electrochemistry,* 43 (2): 119–36. https://doi.org/10.1007/s10800-012-0513-2.

[19] Friedl, J. and Stimming, U. (2013). Model catalyst studies on hydrogen and ethanol oxidation for fuel cells. *Electrochimica Acta,* 101: 41–58. https://doi.org/10.1016/J.ELECTACTA.2012.12.130.

[20] Abdullah, S., Kamarudin, S. K., Hasran, U. A., Masdar, M. S. and Daud, W. R. W. (2014). Modeling and simulation of a direct ethanol fuel cell: An overview. *Journal of Power Sources,* 262: 401–6. https://doi.org/10.1016/J.JPOWSOUR.2014.03.105.

[21] Badwal, S. P. S., Giddey, S., Kulkarni, A., Goel, J. and Basu, S. (2015). Direct ethanol fuel cells for transport and stationary applications – A comprehensive review. *Applied Energy,* 145: 80–103. https://doi.org/10.1016/J.APENERGY.2015.02.002.

[22] Akhairi, M. A. F. and Kamarudin, S. K. (2016). Catalysts in direct ethanol fuel cell (DEFC): An overview. *International Journal of Hydrogen Energy,* 41 (7): 4214–28. https://doi.org/10.1016/J.IJHYDENE.2015.12.145.

[23] Ong, B. C., Kamarudin, S. K. and Basri, S. (2017). Direct liquid fuel cells: A review. *International Journal of Hydrogen Energy,* 42 (15): 10142–57. https://doi.org/10.1016/J.IJHYDENE.2017.01.117.

[24] An, L., Zhao, T. S., Li, Y. S. (2015). Carbon-neutral sustainable energy technology: Direct ethanol fuel cells. *Renewable and Sustainable Energy Reviews,* 50: 1462–68. https://doi.org/10.1016/j.rser.2015.05.074.

[25] Zakil, F. A., Kamarudin, S. K. and Basri, S. (2016). Modified Nafion membranes for direct alcohol fuel cells: An overview. *Renewable and Sustainable Energy Reviews,* 65: 841–52. https://doi.org/10.1016/J.RSER.2016.07.040.

[26] An, L. and Zhao, T. S. (2017). Transport phenomena in alkaline direct ethanol fuel cells for sustainable energy production. *Journal of Power Sources,* 341: 199–211. https://doi.org/10.1016/J.JPOWSOUR.2016.11.117.

[27] Zakaria, Z., Kamarudin, S. K. and Timmiati, S. N. (2016). Membranes for direct ethanol fuel cells: An overview. *Applied Energy,* 163: 334–42. https://doi.org/10.1016/J.APENERGY.2015.10.124.

[28] Sánchez-Monreal, J., Vera, M. and García-Ssalaberri, P. A. (2018). Fundamentals of Electrochemistry with Application to Direct Alcohol Fuel Cell Modeling. Proton Exchange Membrane Fuel Cell. *IntechOpen,* 7: 121–54. http://dx.doi.org/10.5772/intechopen.71635.

[29] Andreadis, G. and Tsiakaras, P. (2006). Ethanol crossover and direct ethanol PEM fuel cell performance modeling and experimental validation. *Chemical Engineering Science,* 61 (22): 7497–508. https://doi.org/10.1016/J.CES.2006.08.028.

[30] Azam, A. M. I. N., Lee, S. H., Masdar, M. S., Zainoodin, A. M. and Kamarudin, S. K. (2018). Parametric study on direct ethanol fuel cell (DEFC) performance and fuel crossover. *International Journal of Hydrogen Energy*, in press. https://doi.org/10.1016/J.IJHYDENE.2018.08.121.

[31] Pramanik, H., Wragg, A. A. and Basu, S. (2008). Studies of some operating parameters and cyclic voltammetry for a direct ethanol fuel cell. *Journal of Applied Electrochemistry,* 38 (9): 1321–28. https://doi.org/10.1007/s10800-008-9560-0.

[32] Andreadis, G. M., Podias, A. K. M. and Tsiakaras, P. E. (2009). A model-based parametric analysis of a direct ethanol polymer electrolyte membrane fuel cell performance. *Journal of Power Sources,* 194 (1): 397–407. https://doi.org/10.1016/J.JPOWSOUR.2009.04.064.

[33] Pramanik, H. and Basu, S. (2010). Modeling and experimental validation of overpotentials of a direct ethanol fuel cell. *Chemical Engineering and Processing: Process Intensification,* 49 (7): 635–42. https://doi.org/10.1016/J.CEP.2009.10.015.

[34] Katayanagi, Y. and Yamazaki, Y. (2010). Effect of Fuel Flow Rate, Fuel Concentration, and O_2 Concentration of the Cathode Gas on the CO_2 Evolution of a Direct Ethanol Fuel Cell. *Electrochemistry,* 12: 976–81.

[35] Alzate, V., Fatih, K. and Wang, H. (2011). Effect of operating parameters and anode diffusion layer on the direct ethanol fuel cell performance. *Journal of Power Sources,* 196 (24): 10625–31. https://doi.org/10.1016/J.JPOWSOUR.2011.08.080.

[36] Heysiattalab, S., Shakeri, M., Safari, M. and Keikha, M. M. (2011). Investigation of key parameters influence on performance of direct ethanol fuel cell (DEFC). *Journal of Industrial and Engineering*

Chemistry, 17 (4): 727–29. https://doi.org/10.1016/J.JIEC.2011.05.037.

[37] Suresh, N. S. and Jayanti, S. (2011). Crossover and performance modeling of liquid-feed Polymer Electrolyte Membrane Direct Ethanol Fuel Cells. *International Journal of Hydrogen Energy,* 36 (22): 14648–58. https://doi.org/10.1016/J.IJHYDENE.2011.07.105.

[38] Naughton, M. S., Tornow, C. E., Bonita, Y., Jhong, H-R "Molly", Brushett, F. R., Gewirth, A. A. and Kenis, P. J. A. (2013). In-situ measurement of ethanol tolerance in an operating fuel cell. *International Journal of Hydrogen Energy,* 38 (21): 8980–91. https://doi.org/10.1016/J.IJHYDENE.2013.04.147.

[39] Chu, Y. H. and Shul, Y. G. (2012). Alcohol Crossover Behavior in Direct Alcohol Fuel Cells (DAFCs) System. *Fuel Cells,* 12 (1): 109–15. https://doi.org/10.1002/fuce.201100044.

[40] Meyer, M., Melke, J. and Gerteisen, D. (2011). Modelling and simulation of a direct ethanol fuel cell considering multistep electrochemical reactions, transport processes and mixed potentials. *Electrochimica Acta,* 56 (11): 4299–4307. https://doi.org/10.1016/J.ELECTACTA.2011.01.070.

[41] James, D. D. and Pickup, P. G. (2010). Effects of crossover on product yields measured for direct ethanol fuel cells. *Electrochimica Acta,* 55 (11): 3824–29. https://doi.org/10.1016/J.ELECTACTA.2010.02.007.

[42] James, D. D. and Pickup, P. G. (2012). Measurement of carbon dioxide yields for ethanol oxidation by operation of a direct ethanol fuel cell in crossover mode. *Electrochimica Acta,* 78: 274–78. https://doi.org/10.1016/J.ELECTACTA.2012.05.120.

[43] Ghumman, A. and Pickup, P. G. (2008). Efficient electrochemical oxidation of ethanol to carbon dioxide in a fuel cell at ambient temperature. *Journal of Power Sources,* 179 (1): 280–85. https://doi.org/10.1016/J.JPOWSOUR.2007.12.071.

[44] Jablonski, A., Kulesza, P. J. and Lewera, A. (2011). Oxygen permeation through Nafion 117 membrane and its impact on efficiency of polymer membrane ethanol fuel cell. *Journal of Power Sources,* 196 (10): 4714–18. https://doi.org/10.1016/J.JPOWSOUR.2011.01.045.

[45] Fu, R. Q., Hong, L. and Lee, J. Y. (2008). Membrane Design for Direct Ethanol Fuel Cells: A Hybrid Proton-Conducting Interpenetrating Polymer Network. *Fuel Cells,* 8 (1): 52–61. https://doi.org/10.1002/fuce.200700041.

[46] Barbora, L., Acharya, S. and Verma, A. (2009). Synthesis and Ex-situ Characterization of Nafion/TiO_2 Composite Membranes for Direct Ethanol Fuel Cell. *Macromolecular Symposia,* 277 (1): 177–89. https://doi.org/10.1002/masy.200950322.

[47] Roelofs, K. S., Hirth, T. and Schiestel, T. (2010). Sulfonated poly(ether ether ketone)-based silica nanocomposite membranes for direct ethanol fuel cells. *Journal of Membrane Science,* 346 (1): 215–26. https://doi.org/10.1016/J.MEMSCI.2009.09.041.

[48] Roelofs, K. S. and Schiestel, T. (2010). sPEEK based composite membranes for direct ethanol fuel cell applications. *Desalination,* 250 (3): 1051–52. https://doi.org/10.1016/J.DESAL.2009.09.105.

[49] Maab, H. and Nunes, S. P. (2010). Modified SPEEK membranes for direct ethanol fuel cell. *Journal of Power Sources,* 195 (13): 4036–42. https://doi.org/10.1016/J.JPOWSOUR.2010.01.005.

[50] Roelofs, K. S., Hirth, T. and Schiestel, T. (2011). Dihydrogenimidazole modified silica-sulfonated poly(ether ether ketone) hybrid materials as electrolyte membranes for direct ethanol fuel cells. *Materials Science and Engineering: B,* 176 (9): 727–35. https://doi.org/10.1016/J.MSEB.2011.02.029.

[51] Kawaguti, C. A., Dahmouche, K. and Gomes, A. S. (2012). Nanostructure and properties of proton-conducting sulfonated poly(ether ether ketone) (SPEEK) and zirconia-SPEEK hybrid membranes for direct alcohol fuel cells: effect of the nature of swelling solvent and incorporation of heteropolyacid. *Polymer International,* 61 (1): 82–92. https://doi.org/10.1002/pi.3151.

[52] Tang, H., Pan, M. and Jiang, S. P. (2011). Self assembled 12-tungstophosphoric acid–silica mesoporous nanocomposites as proton exchange membranes for direct alcohol fuel cells. *Dalton Transactions,* 40 (19): 5220-27. https://doi.org/10.1039/c1dt10150a.

[53] Kumar, G. G. (2011). Irradiated PVdF-HFP-montmorillonite composite membranes for the application of direct ethanol fuel cells. *Journal of Materials Chemistry,* 21: 17382-91. https://doi.org/10.1039/c1jm11528f.

[54] Gomes, A. S. and Filho, J. C. D. (2012). Hybrid membranes of PVA for direct ethanol fuel cells (DEFCs) applications. *International Journal of Hydrogen Energy,* 37 (7): 6246–52. https://doi.org/10.1016/J.IJHYDENE.2011.08.002.

[55] Kim, I., Han, O. H., Chae, S. A., Paik, Y., Kwon, S.-H., Lee, K.-S., Sung, Y.-E. and Kim, H. (2011). Catalytic Reactions in Direct Ethanol Fuel Cells. *Angewandte Chemie,* 123 (10): 2318–22. https://doi.org/10.1002/ange.201005745.

[56] Biswas, S. K., Sambu, P. and Basu, S. (2009). Influence of a pore-former and PTFE in the performance of the direct ethanol fuel cell. *Asia-Pacific Journal of Chemical Engineering,* 4 (1): 3–7. https://doi.org/10.1002/apj.192.

[57] Beyhan, S., Coutanceau, C., Léger, J.-M., Napporn, T. W. and Kadırgan, F. (2013). Promising anode candidates for direct ethanol fuel cell: Carbon supported PtSn-based trimetallic catalysts prepared by Bönnemann method. *International Journal of Hydrogen Energy,* 38 (16): 6830–41. https://doi.org/10.1016/j.ijhydene.2013.03.058.

[58] Basu, S., Agarwal, A. and Pramanik, H. (2008). Improvement in performance of a direct ethanol fuel cell: Effect of sulfuric acid and Ni-mesh. *Electrochemistry Communications,* 10 (9): 1254–57. https://doi.org/10.1016/J.ELECOM.2008.05.042.

[59] Wang, Q., Sun, G. Q., Cao, L., Jiang, L. H., Wang, G. X., Wang, S. L., Yang, S. H. and Xin, Q. (2008). High performance direct ethanol fuel cell with double-layered anode catalyst layer. *Journal of Power Sources,* 177 (1): 142–47. https://doi.org/10.1016/J.JPOWSOUR.2007.11.040.

[60] Wan, C.-H. and Chen, C.-L. (2009). Mitigating ethanol crossover in DEFC: A composite anode with a thin layer of Pt50–Sn50 nanoparticles directly deposited into Nafion® membrane surface.

International Journal of Hydrogen Energy, 34 (23): 9515–22. https://doi.org/10.1016/J.IJHYDENE.2009.09.075.

[61] Ekdharmasuit, P., Therdthianwong, A. and Therdthianwong, S. (2013). Anode structure design for generating high stable power output for direct ethanol fuel cells. *Fuel,* 113: 69–76. https://doi.org/10.1016/J.FUEL.2013.05.046.

[62] Sarris, I., Tsiakaras, P., Song, S. and Vlachos, N. (2006). A three-dimensional CFD model of direct ethanol fuel cells: Anode flow bed analysis. *Solid State Ionics,* 177 (19–25): 2133–38. https://doi.org/10.1016/J.SSI.2006.02.019.

[63] Orozco, D. C., Sánchez, C., Lobato, J. and Úbeda, D. (2011). Hydrodynamics and Current Distribution Analysis of Bipolar Plates for Direct Ethanol Fuel Cells. *ECS Transactions - The Electrochemical Society,* 41: 1909–25. https://doi.org/10.1149/1.3635721.

[64] Oliveira, V. B., Falcão, D. S., Rangel, C. M. and Pinto, A. M. F. R. (2007). A comparative study of approaches to direct methanol fuel cells modelling. *International Journal of Hydrogen Energy,* 32 (3): 415–24. https://doi.org/10.1016/J.IJHYDENE.2006.06.049.

[65] Andreadis, G. M., Podias, A. K. M. and Tsiakaras, P. E. (2008). The effect of the parasitic current on the Direct Ethanol PEM Fuel Cell Operation. *Journal of Power Sources,* 181 (2): 214–27. https://doi.org/10.1016/J.JPOWSOUR.2008.01.060.

[66] Andreadis, G., Song, S. and Tsiakaras, P. (2006). Direct ethanol fuel cell anode simulation model. *Journal of Power Sources,* 157 (2): 657–65. https://doi.org/10.1016/J.JPOWSOUR.2005.12.040.

[67] Sousa, R., Anjos, D. M., Tremiliosi-Filho, G., Gonzalez, E. R., Coutanceau, C., Sibert, E., Léger, J.-M. and Kokoh, K. B. (2008). Modeling and simulation of the anode in direct ethanol fuels cells. *Journal of Power Sources,* 180 (1): 283–93. https://doi.org/10.1016/J.JPOWSOUR.2008.01.058.

[68] Goel, J. and Basu, S. (2015). Mathematical modeling and experimental validation of direct ethanol fuel cell. *International Journal of Hydrogen Energy,* 40 (41): 14405–15. https://doi.org/10.1016/J.IJHYDENE.2015.03.082.

[69] Souza, M. M., Gomes, R. S. and Bortoli, A. L. (2018). A model for direct ethanol fuel cells considering variations in the concentration of the species. *International Journal of Hydrogen Energy,* 43 (29): 13475–88. https://doi.org/10.1016/J.IJHYDENE.2018.05.096.

[70] Pinto, A. M. F. R., Oliveira, V. B. and Falcão, D. S. (2018). *Direct Alcohol Fuel Cells for Portable Applications: Fundamentals, Engineering and Advances*. Academic Press Elsevier, 1-353.

[71] Pereira, J. P., Falcão, D. S., Oliveira, V. B. and Pinto, A. M. F. R. (2014). Performance of a passive direct ethanol fuel cell. *Journal of Power Sources,* 256: 14–19. https://doi.org/10.1016/J.JPOWSOUR.2013.12.036.

[72] Oliveira, V. B., Pereira, J. P. and Pinto, A. M. F. R. (2017). Modeling of passive direct ethanol fuel cells. *Energy,* 133: 652–65. https://doi.org/10.1016/J.ENERGY.2017.05.152.

In: Advances in Energy Research. Volume 31 ISBN: 978-1-53616-271-4
Editor: Morena J. Acosta © 2019 Nova Science Publishers, Inc.

Chapter 4

PERFORMANCE OF A PASSIVE DIRECT ALCOHOL FUEL CELL FED WITH A MIXTURE OF ALCOHOLS

B. A. Braz, C. S. Moreira, V. B. Oliveira [*]
and A. M. F. R. Pinto [†]
Transport Phenomena Research Center (CEFT),
Department of Chemical Engineering,
Faculty of Engineering University of Porto, Porto, Portugal

ABSTRACT

In recent years, the demand for efficient, renewable and sustainable energy sources has been increasing in order to obtain a more diverse energy production. Direct alcohol fuel cells (DAFCs) appear as a very promising power solution for portable devices, since they offer high energy densities and longer runtime than batteries, instant recharging, use a liquid fuel (alcohol), operate at room temperature and can rely on a passive flow of reactants and products, allowing to operate the cell without any additional

[*] Corresponding Author's E-mail: vaniaso@fe.up.pt.
[†] Corresponding Author's E-mail: apinto@fe.up.pt.

power consumption. Therefore, much work has been done towards the optimization of passive DAFC systems, where the fuel and the oxidant are feed in a passive mode, and some progress towards its commercialization has already been achieved. Despite that, these systems still have costs above the desirable and a lower lifetime and power output.

The passive DAFCs can be supplied with different alcohols, but the most used and studied ones are methanol and ethanol. Methanol has the highest energy to carbon ratio of any other alcohol, but is toxic for humans and is non-renewable. Ethanol emerges as an alternative and promising fuel due to its non-toxicity, natural availability and renewability. Nevertheless, the performance of direct ethanol fuel cells (DEFCs) is lower than the DMFCs (direct methanol fuel cells), due to the sluggish ethanol electrochemical oxidation, since it is very difficult to cleavage the carbon-carbon bond presented on the ethanol structure. A novel idea to achieve higher performances with passive DAFCs is to use a mixture of methanol and ethanol, since it is expected to obtain the best characteristics of each alcohol. In order to achieve that, the present chapter presents a study with a passive DAFC fed with a mixture of methanol and ethanol with different concentrations and volume ratios, towards its optimization. The experimental results were analyzed under the light of the polarization and electrochemical impedance spectroscopy (EIS) measurements. The EIS technique was used to identify and quantify the different losses that negatively affect the passive DAFC performance, pointing out possible solutions to minimize them.

Keywords: direct alcohol fuel cell, passive, methanol, ethanol, electrochemical impedance spectroscopy, performance

INTRODUCTION

The massive use of new portable power devices and the conception of decentralized grid-independent systems are part of the solution to the energy problem. Fuel Cells (FCs) are in the Energy Agenda and have market potential in education, recreational and military applications and in auxiliary power systems and portable devices. Compact direct alcohol fuel cells (DAFCs) using alcohols as fuel without a reforming step appear as a good solution to develop advanced portable FCs to overcome systematic limitations of conventional batteries (Wongyao, Therdthianwong and

Therdthianwong 2011; Ong, Kamarudin and Basri 2017; Zakil, Kamarudin and Basri 2016; Gupta and Pramanik 2019).

In a DAFC, fuel and oxidant can be supplied by a passive mode where natural transport mechanisms, such as diffusion, convection and evaporation are used to achieve all the processes without any additional power consumption. Therefore, these systems are compact and simpler, have a fast refueling, operate at ambient conditions and have lower cost than the active ones. These advantages make this technology attractive to the rapid growing need for portable power sources. However, the disadvantages of using DAFCs are the slow anode kinetics, resulting in higher anodic overpotentials and the alcohol crossover, which not only lowers the fuel utilization but also degrades the cathode performance and generates extra heat, higher costs and a lower durability. Among all the possible fuels for these systems, liquid alcohol fuels, which have high energy density and are easy to transport, store and handle become an alternative to hydrogen. Up to a few years ago, methanol was the most used fuel and much progress was made in the development and optimization of passive direct methanol fuel cells (DMFCs). However, methanol is toxic for human beings, is easily volatile and inflammable as well as nonrenewable (Wongyao, Therdthianwong and Therdthianwong 2011). Additionally, the use of methanol as fuel leads to higher rates of methanol crossover through the membrane towards the cathode side, due to its lower molecular size, decreasing the fuel cell performance due to the formation of a mixed potential at the cathode side and a loss of the available fuel at the anode side. The mixed potential occurs due to an additional electrochemical reaction at the cathode side, the methanol oxidation, which not only lowers the catalyst available sites for the oxygen reduction reaction but also poisons the cathode catalyst with the intermediate compounds formed due to incomplete methanol oxidation reaction on the Pt catalyst (Xu, Faghri and Li 2010; Park et al. 2012; Yuan et al. 2012; Oliveira, Pereira and Pinto 2016; Braz et al. 2019).

Among the other available alcohols, ethanol appears as an attractive and promising fuel since is non-toxic, has a natural availability and renewability and a higher energy density than methanol. Moreover, ethanol has a lower crossover rate than methanol mainly due to its larger molecular size and thus its effect on the cathode performance and on the overall cell performance is less severe (Wongyao, Therdthianwong and Therdthianwong 2011; Pereira et al. 2014; S. Song et al. 2005; Badwal et al. 2015; Leo et al. 2013). However, the cleavage of the carbon-carbon bond, of the ethanol molecule, is very difficult making the ethanol oxidation rate rather low. Platinum (Pt) is the most active catalyst used, however if used alone, self-inhibition occurs. An additive such as Tin (Sn) and Ruthenium (Ru) is used to modify the Pt surface. Nevertheless, the DEFC (direct ethanol fuel cells) performance is still poor, due to the incomplete oxidation of ethanol and needs to be improved.

An innovative idea is to explore the possibility of achieving higher performances operating the cells with a mixture of methanol and ethanol, which is expected to show the best advantages of using both alcohols and is a very promising but still not explored research field. Therefore, in this work the effect of the methanol/ethanol ratio on the performance of a pDAFC using different fuel concentrations is evaluated. The cell behavior was analyzed through polarization and electrochemical impedance spectroscopy (EIS) measurements. The EIS technique allows the identification and quantification of the mains losses that negatively affect the fuel cell performance, highlighting out possible measures and solutions to avoid or decrease them (Piela, Fields and Zelenay 2006; Braz et al. 2019).

MATERIALS AND METHODS

This section presents a description of the passive DAFC used in the experimental studies, as well as the different materials and conditions tested. Furthermore, the features of the test station are presented and the experimental procedures adopted, regarding the polarization and the EIS measurements and electrical circuit fitting, are described.

Fuel Cell Design and Conditions

The pDAFC consists of two acrylic end plates (open at the cathode side and with a fuel reservoir at the anode side), two isolating rubber plates, two stainless steel current collector plates (with a free open ratio of 41%), two diffusion layers (DLs), two catalyst layers (CLs) and a polymer membrane between the anode and the cathode (Figure 1).

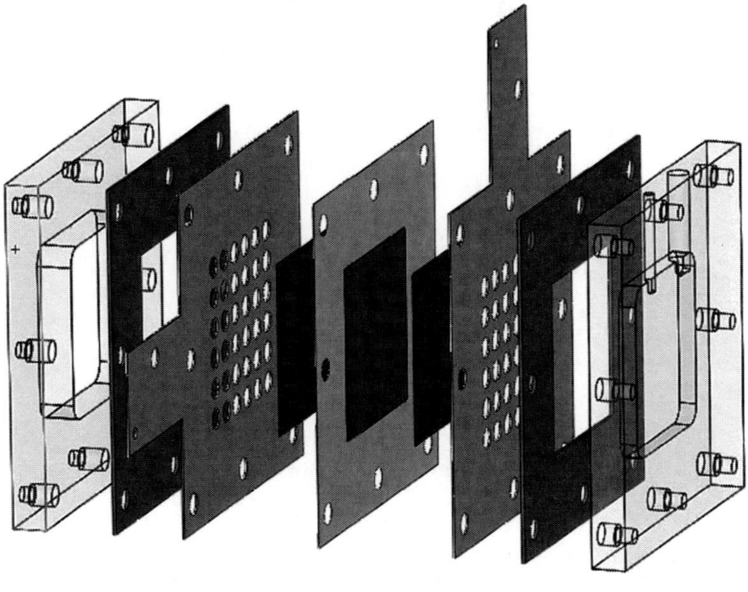

Figure 1. Schematic representation of the "in-house" passive DAFC.

The pDAFC used has an active area of 25 cm^2 and was operated at ambient conditions (atmospheric pressure and ambient temperature). The cell was operated with two different MEAs, named as commercial and tailored MEA. The commercial MEA is a five-layer MEA, acquired from QuinTech, and is composed by a Nafion 117 membrane, 3 mg/cm^2 of Pt/Ru at the anode side and 1.2-1.4 mg/cm^2 of Pt at the cathode side and Freudenberg H2315 I2C6 as anode DL and Freudenberg H2315 T10A as cathode DL (Table 1). The tailored MEA consists on a Nafion 117 membrane, 3 mg/cm^2 of Pt/Ru as anode catalyst and 1.2-1.4 mg/cm^2 of Pt as cathode catalyst, acquired from QuinTech, and carbon cloth with a microporous layer (CC_MPL) as anode DL and carbon cloth (CC) as cathode DL (Table 1). The tailored MEA was pressed by a non-bonded mode when assembling the fuel cell.

Table 1. Properties of the materials used as diffusion layers

Diffusion layer	Type	Thickness (mm)	MPL	PTFE treatment
CC	Cloth	0.400	No	No
CC_MPL	Cloth	0.410	Yes	No
Freudenberg H2315 I2C6	Paper	0.250	Yes	Yes
Freudenberg H2315 T10A	Paper	0.215	No	Yes

Table 2. Methanol/ethanol ratios tested for the different fuel concentrations

Methanol/ethanol ratio	Methanol volume (%)	Ethanol volume (%)
100:0	100	0
75:25	75	25
50:50	50	50
25:75	25	75
0:100	0	100

The experimental tests were performed with different fuel concentrations, 2 M, 3 M and 5 M, and different methanol/ethanol ratios, as can be seen in Table 2. The methanol used had a purity of 99.99% and was supplied by Fischer Chemical. Ethanol had a minimum purity of 99.5% and was supplied by AGA.

Electrochemical Measurements

The electrochemical measurements were performed with an electrochemical test station, Zahner Elektrik GmbH &Co. KG. The polarization measurements were conducted galvanostatically and the impedance ones potentiostatically. The EIS tests were carried out with an amplitude of 10 mV, a frequency range from 10 mHz to 100 kHz and at different fuel cell voltages, 0.45 V, 0.4 V, 0.35 V, 0.3 V and 0.2 V.

The EIS technique was performed in-situ to allow the measurements of the fuel cell as a whole. This is the common procedure used to analyze a single fuel cell, since the information provided by this approach helps on the optimization of the design parameters and operating conditions. As the use of a reference electrode in most of the fuel cell systems is difficult and unpractical, in this work a dynamic hydrogen electrode (DHE) was used, by replacing the methanol or air supply by hydrogen. With this approach, it is assumed that the losses associated with the DHE are irrelevant and consequently all the measurements are attributed to the other electrode, the electrode under study. Therefore, when the DHE is used at the cathode, it is possible to study the anode behavior and obtain the anode spectrum (Figure 2) and when the DHE is used at the anode, it is possible to study the cathode behavior and obtain the cathode spectrum (Figure 3). Then, performing the impedance measurements with the pDAFC working with methanol and air, it is possible to obtain the overall pDAFC spectrum (Figure 4). Comparing it with the single-cell ones allows distinguishing the different electrode losses (anode and cathode) on the overall cell response (Piela, Fields and Zelenay 2006; Braz et al. 2019). The impedance spectra was, then, fitted to an equivalent electric circuit (EEC) (Figure 5), using the Thales software from Zahner, to quantify the different losses that affect the cell performance (Braz et al. 2019).

RESULTS AND DISCUSSION

This section presents the experimental results obtained with the "in-house" pDAFC, which were explained under the light of the EIS data. This section starts with the interpretation of the EIS data using a selected set of values and the EEC quality fitting. Then, the effect of using different methanol and ethanol ratios for different fuel concentrations and two different MEAs on the passive DAFC behavior is discussed. An economic evaluation where the tailored MEA is compared with the commercial one is presented at the end of the section.

EIS Measurements and EEC Fitting

The results presented in this section were obtained with the tailored MEA and with methanol as fuel with a concentration of 1 M.

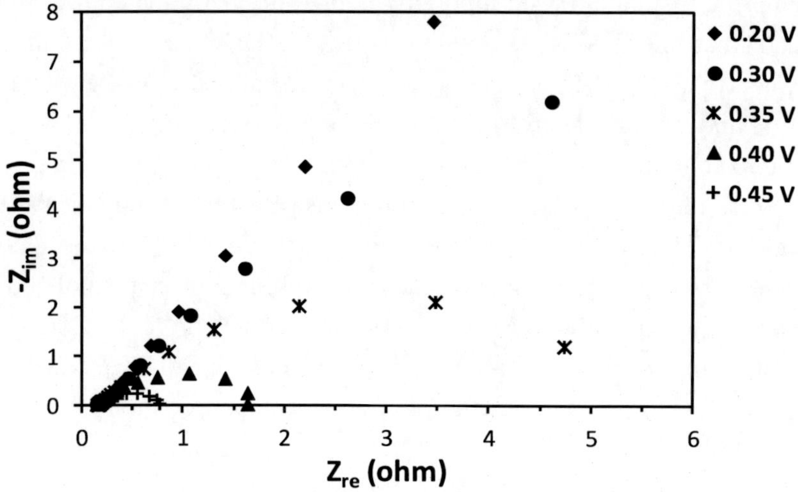

Figure 2. Nyquist plot of a pDAFC with the DHE at the cathode side for different cell voltages.

As already mentioned, the EIS measurements can provide detailed information about the fuel cell system, making possible the identification and quantification of the different voltage losses that negatively affect the fuel cell performance: ohmic, activation and mass transport losses. The EIS data are usually represented by a Nyquist plot, such as the ones shown in Figures 2, 3 and 4, where the real impedance (Z_{re}) is plotted against the imaginary one (Z_{im}). The characteristic and shape of each spectrum make possible identifying the different losses that affect the cell performance. Therefore, different combinations of several electrical elements have to be considered on the EEC model that will be used to model the impedance spectrum (Piela, Fields and Zelenay 2006; Braz et al. 2019).

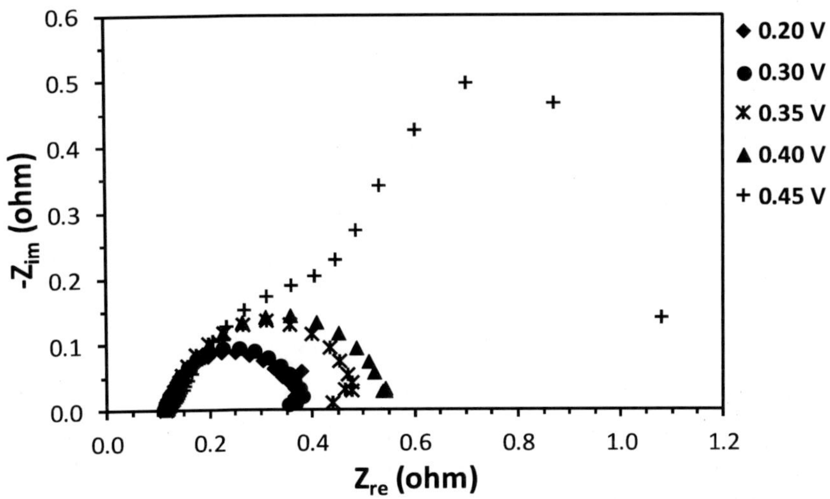

Figure 3. Nyquist plot of a pDAFC with the DHE at the anode side for different cell voltages.

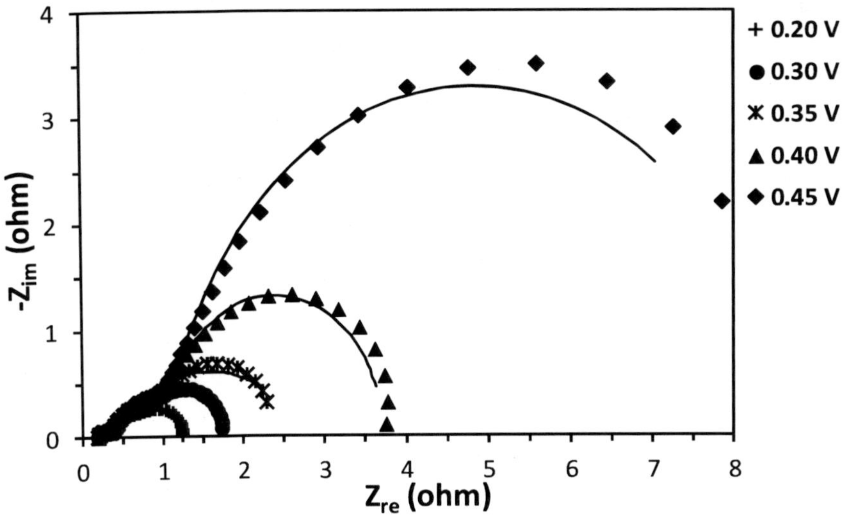

Figure 4. Nyquist plot of a pDAFC fed with methanol and air for different cell voltages.

A typical Nyquist plot of a fuel cell can be divided into different regions according to its frequency range. At the high frequencies region, the imaginary impedance axis intercepts the real impedance one at a value that represents the ohmic losses (as shown in Figures 2, 3 and 4). This point is represented in an electrical circuit by a resistor (R) (R_{Ohm} in the EEC shown in Figure 5) and due to its frequency independency, its impedance has only a real part, identified by a single point at the real axis (X-axis) of the Nyquist plot, with a value equal to R or R_{Ohm}.

The arc presented at the medium frequency region represents the activation losses and is characterized by a decrease of the resistance with a decrease of the voltage (as shown in Figures 2, 3 and 4 and Table 3) (Piela, Fields and Zelenay 2006, Braz et al. 2019). The arc at the lower frequency region, usually, represents the mass transport losses and is characterized by an increase of the resistance with a decrease of voltage (Piela, Fields and Zelenay 2006). However, in some systems, as the one under study, it is not possible to reach the mass transport losses region and therefore the arcs, at the medium and lower frequency ranges, are due to the activation losses, characterized by a decrease of the resistance with a decrease of the voltage (as shown in Figure 3 and 4 and Table 3) (Braz et al. 2019). This allows to conclude that under the operating conditions studied in this work, the pDAFC does not suffer from concentration or mass transport losses and the activation ones are dominant for all the voltages studied. As known the activation losses are due to the electrochemical reactions that occur in a pDAFC system, the alcohol oxidation and the oxygen reduction, which justify two of the arcs presented in Figure 4. The other arc, the third one, is characteristic of a DAFC and it is due to an additional electrochemical reaction that occurs at the cathode side, the alcohol oxidation due to alcohol crossover (Braz et al. 2019). This additional resistance that affects mainly the cathode side, can be clearly identified on the cathode spectrum, shown in Figure 3, where both resistances decrease with a decrease of the cell voltage, a normal behavior of the activation losses. Therefore, both resistances presented on the cathode spectrum are due to the two reactions that occur at this side. Additionally, it is well known that the crossover decreases with a decrease of voltage (Oliveira, Rangel and Pinto 2011), and accordingly, the crossover effect, here estimated by the $R_{Crossover}$, decreases with a decrease of the voltage, as shown in Figures 3 and 4 and Table 3.

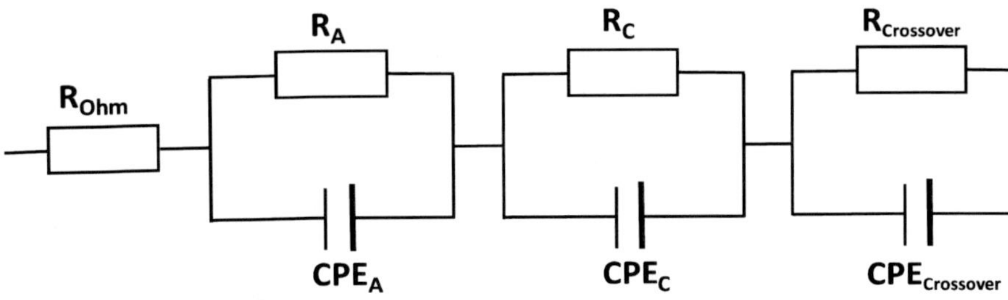

Figure 5. Equivalent electric circuit (EEC).

Table 3. Values of the different resistances obtained by fitting the Nyquist plots of a pDAFC (Figure 4) with the EEC (Figure 5) for different fuel cell voltages

Voltage (V)	R_{Ohm} (Ω)	R_A (Ω)	R_C (Ω)	$R_{Crossover}$ (Ω)
0.45	0.18	0.15	0.47	7.2
0.40	0.18	0.15	0.47	3.1
0.35	0.18	0.15	0.47	1.5
0.30	0.18	0.15	0.32	1.2
0.20	0.18	0.15	0.17	0.7

Based on the EIS measurements and its interpretation, the EEC used in this work, consists on a resistance, R_{Ohm} (ohmic resistance) connected in series with other three independent circuits. Each one comprising a resistance, R_A (activation losses due to the methanol oxidation at the anode), R_C (activation losses due to the oxygen reduction at the cathode) and $R_{Crossover}$ (activation losses due to the methanol oxidation at the cathode) in parallel with a constant phase element, CPE_A, CPE_C and $CPE_{Crossover}$, associated with the capacitance properties of the double-layer interfaces (Figure 5) (Braz et al. 2019).

From Figure 4 it is possible to see a good agreement between the EEC fitting and the experimental results, revealing that the EEC reproduces with accuracy the system under study.

It should be here highlighted, that the impedance spectra obtained for the other conditions studied followed the same pattern as the one presented in this section, with similar results for the EEC fitting and resistance values. Therefore, the EIS data for a voltage of 0.2 V, near the voltage corresponding to the maximum power density for each condition tested, was selected to be presented in the following sections and will be used to explain the pDAFC behavior.

Effect of Methanol/Ethanol Ratio Using a Tailored MEA

In this section, a tailored MEA was used to study the effect of the methanol/ethanol ratio on the pDAFC behavior for different fuel concentrations, 2 M, 3 M and 5 M. The polarization measurements are presented in Figure 6 and the values for the different resistances of the EEC, at 0.2 V, and the maximum power density achieved for each condition tested can be found in Table 4.

As can be seen in Figure 6 and Table 4, the best results using the tailored MEA were achieved with a lower amount of ethanol, lower volume ratio, for all the three fuel concentrations tested, 2 M, 3 M and 5 M. However, for 2 M similar performances were achieved with an ethanol ratio of 25%, 50% and 75%. These results can be explained by the fact that the anode activation losses increase with an increase of the ethanol ratio due to its slow reaction rate, since the catalyst used, in this work, Pt/Ru, is not the adequate one for this reaction and to the misplaced adsorption of ethanol on Pt/Ru catalysts (Wongyao, Therdthianwong and Therdthianwong 2011). Additionally, a higher ethanol ratio will lead to a lower anode catalyst activity due to its poisoning by the by-products from the incomplete ethanol oxidation reaction, leading to a decrease of the active sites for the methanol oxidation reaction. Therefore, less methanol reacts on the anode side and more methanol reaches the membrane and crosses it towards the cathode side (higher $R_{Crossover}$). This will increase the cathode activation losses due to cathode poisoning by the methanol oxidation at this side, which will decrease the oxygen reduction rate and therefore increase the cathode activation losses due to this reaction (R_C) and to the undesired one ($R_{Crossover}$). As expected, and as can be seen by the resistance values presented in Table 4, the crossover effect ($R_{Crossover}$) is more severe for higher ethanol ratios and fuel concentrations (Oliveira, Rangel and Pinto 2011).

Table 4. Values for the different resistances of the EEC for different methanol/ethanol ratios and fuel concentrations and its maximum power density

Fuel Concentration	Methanol/Ethanol Ratio	R_{Ohm} (Ω)	R_A (Ω)	R_C (Ω)	$R_{Crossover}$ (Ω)	Maximum Power density (mW/cm^2)
2 M	100:0	0.46	0.02	0.30	0.47	1.43
	75:25	0.29	0.02	0.53	2.70	0.39
	50:50	0.30	0.02	0.68	3.95	0.35
	25:75	0.35	0.02	0.83	5.93	0.29
	0:100	0.33	0.02	1.20	9.68	0.23
3 M	100:0	0.42	0.03	0.28	0.30	1.66
	75:25	0.29	0.02	0.52	2.07	0.55
	50:50	0.29	0.03	0.60	4.02	0.38
	25:75	0.30	0.02	0.92	5.95	0.29
	0:100	0.25	0.06	0.76	10.4	0.23
5 M	100:0	0.41	0.02	0.13	0.24	3.00
	75:25	0.57	0.04	0.89	2.67	0.37
	50:50	0.52	0.09	1.73	5.16	0.21
	25:75	0.53	0.10	4.80	18.2	0.08
	0:100	0.54	0.10	8.36	31.3	0.06

For the tailored MEA, the best performance was achieved for a methanol/ethanol ratio of 100:0 and a fuel concentration of 5 M (Figure 6 and Table 4).

For the ethanol oxidation a binary catalyst of Pt/Sn would be more appropriate (Zhou et al. 2004; Song et al. 2005), since it would increase the electrocatalytic activity and the ethanol oxidation rate. In this case, to attend the oxidation requirements of each alcohol, it would be preferable to use a mixed catalyst, which is not commercially available.

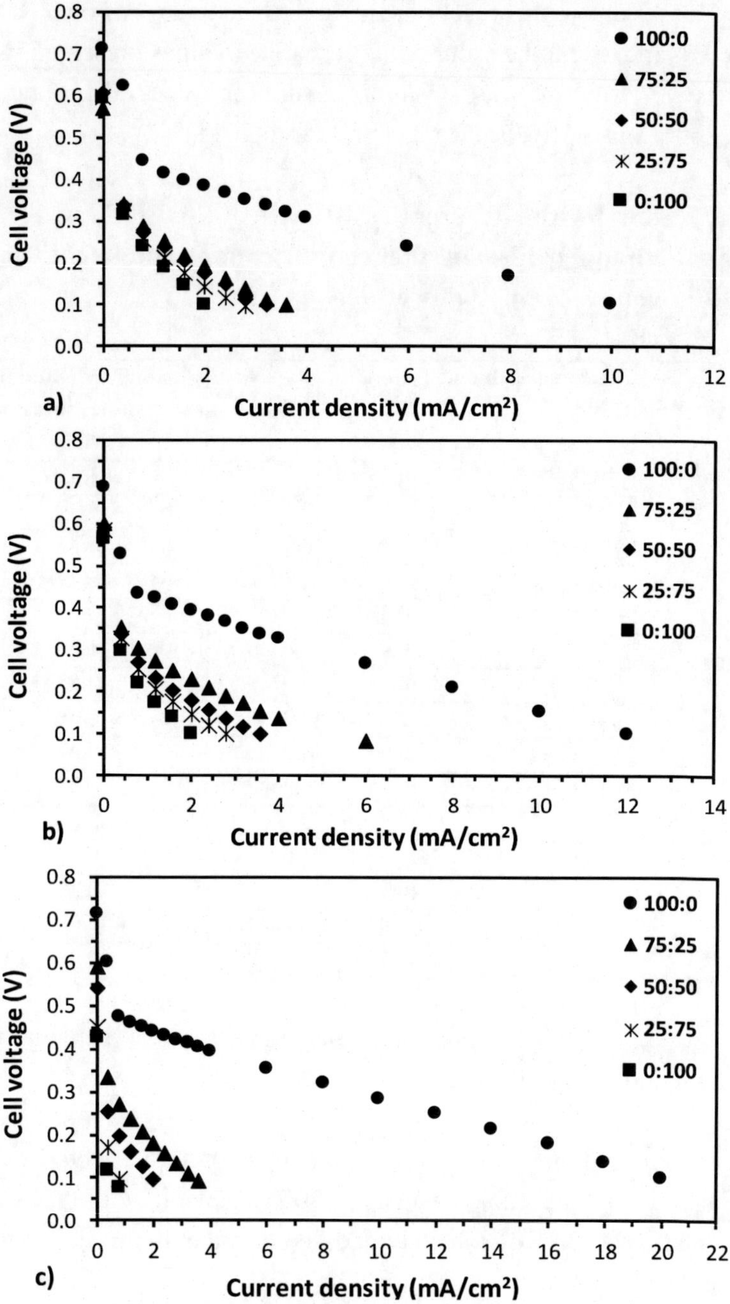

Figure 6. Effect of methanol/ethanol ratio on the pDAFC performance for different fuel concentrations: a) 2 M, b) 3 M and c) 5 M.

Effect of Methanol/Ethanol Ratio Using a Commercial MEA

In this section, the effect of the methanol/ethanol ratio on the pDAFC performance was evaluated using different fuel concentrations and a commercial MEA. Figure 7 shows the polarization curves for each condition tested and in Table 5 can be found the values for the ohmic and activation resistances of the EEC at 0.2 V as well as the cell maximum power density.

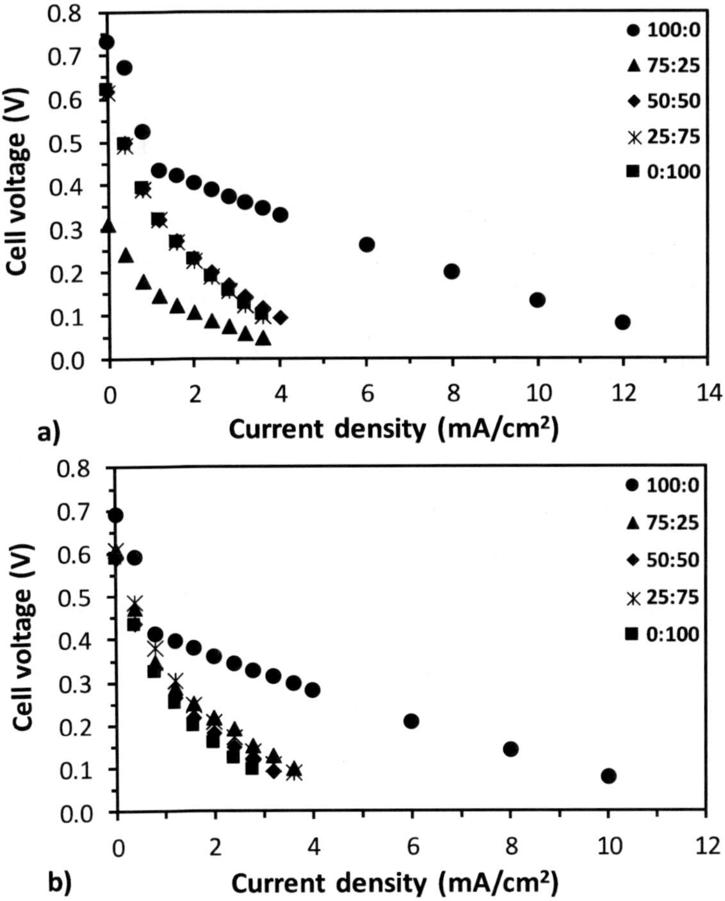

Figure 7. Effect of methanol/ethanol ratio on the pDAFC performance for different fuel concentrations: a) 2 M and b) 3 M.

Table 5. Values for the different resistances of the EEC for different methanol/ethanol ratios and fuel concentrations and its maximum power density

Fuel Concentration	Methanol/Ethanol Ratio	R_{Ohm} (Ω)	R_A (Ω)	R_C (Ω)	$R_{Crossover}$ (Ω)	Maximum Power density (mW/cm^2)
2 M	100:0	0.52	0.04	0.44	0.36	1.58
	75:25	0.71	0.07	0.62	2.38	0.21
	50:50	0.59	0.10	0.90	2.29	0.48
	25:75	0.73	0.09	1.58	2.33	0.45
	0:100	0.48	0.10	1.82	2.27	0.46
3 M	100:0	0.56	0.04	0.44	0.43	1.25
	75:25	0.75	0.12	0.88	2.05	0.46
	50:50	0.63	0.12	0.75	2.37	0.36
	25:75	0.65	0.10	2.25	2.30	0.42
	0:100	0.71	0.15	1.30	2.84	0.32

As happened in the study regarding the effect of the methanol/ethanol ratio on the pDAFC power output using a tailored MEA, the best performance using the commercial MEA was achieved using only methanol as fuel (Figure 7 and Table 5). The decrease of power output with an increase of the ethanol ratio on the fuel mixture can be explained by the lower kinetic rate of the ethanol oxidation reaction at the Pt/Ru sites than the methanol one, due to the cleavage of the strong C-C bond of the ethanol molecule and to the fact that this catalyst is not the most adequate one to promote a complete oxidation of ethanol to carbon dioxide (Ong, Kamarudin and Basri 2017; Pereira et al. 2014). However, for the higher fuel concentration tested, 3 M, similar power outputs were achieved with and a methanol/ethanol ratio of 75:25, 50:50 and 25:75, showing that if the power requirements of the system where the pDAFC will be implemented are not very demanding it is possible to feed the cell with a lower amount of methanol. This behavior can be explained by the similar $R_{Crossover}$ values obtained for these three methanol/ethanol ratios (Table 5). As the fuel crossover is one on the main causes for the loss of performance in these systems, similar crossover rates will lead to similar performances. Despite that, these values are very high, showing that the fuel crosses the membrane towards the cathode side, reacting at this side and decreasing the cathode performance and the cathode catalyst activity for the oxygen reduction reaction, leading to higher cathode activation losses (R_C). As mentioned before, the fuel crossover is a major drawback of DAFCs, and as this phenomenon is directly related with the concentration gradient between the anode and the cathode sides its effect increases with an increase of the fuel concentration. Therefore, as can be seen in Table 5 and as expected, the $R_{Crossover}$ increases with an increase of the fuel concentration.

For the commercial MEA, the best performance was achieved for a methanol/ethanol ratio of 100:0 and a fuel concentration of 2 M (Figure 7 and Table 5). However, a similar power output, 1.25 mW/cm^2 was achieved for a methanol/ethanol ratio of 100:0 and a fuel concentration of 3 M.

ECONOMIC EVALUATION

To reach its commercialization and massive use, the pDAFC systems must be cost competitive and have similar investment and manufacturing costs than the conventional technologies that they will replace. These systems overall costs include material, manufacturing and operating and maintenance costs, being the material ones responsible for the major fraction on the overall system cost (Wang, Wang and Fan 2018). Regarding them, it is known that the major component responsible for its higher value are the MEA.

In the recent years, some efforts have been done to decrease the material costs of pDAFCs by reducing the catalyst loadings, developing new catalyst made from low-cost materials and developing low-cost membranes using low-cost materials for the diffusion layers (Bambagioni et al. 2009; Beyhan et al. 2013; Hu et al. 2013; Zainoodin et al. 2014; Zakaria, Kamarudin and Timmiati 2016; Carvalho, Colmati and Tanaka 2017; Ong, Kamarudin and Basri 2017; Herranz et al. 2018; Braz et al. 2019).

Table 6. Costs of the MEAs used in this work

MEA	Characteristic	Cost (€)
Commercial	Anode 3.0 mg/cm^2 Pt/Ru and Freudenberg H2315 I2C6; Cathode 1.2-1.4 mg/cm^2 Pt and Freudenberg H2315 T10A; Nafion 117	150.00
Tailored	Anode 3.0 mg/cm^2 Pt/Ru and carbon cloth with MPL (CC_MPL); Cathode 1.2-1.4 mg/cm^2 Pt and carbon cloth (CC); Nafion 117	137.83

Having this in mind, one of the goals of this work was to optimize the performance of a pDAFC by testing different MEAs, commercial and tailored, with different properties and costs, and using lower loadings on both anode and cathode catalyst layers, 3.0 mg/cm^2 Pt/Ru and 1.2-1.4 mg/cm^2 Pt, than the conventional ones used for these systems, 4.0 mg/cm^2 Pt/Ru and 4 mg/cm^2 Pt.

Table 6 shows the costs of the two MEAs used in the current work. As can be seen the one proposed in this work, the tailored MEA, allowed a cost reduction of 8% and a higher power output 3.00 mW/cm^2 than the commercial one. However, this value is much lower than the ones already presented in literature and can be explained by the lower loadings used in this work. These was performed in order to decrease the catalyst costs and consequently the overall system costs. Additionally, this low-cost MEA can be used in some particular applications where the power requirements are not very high.

CONCLUSION

Methanol has been considered the most promising alcohol to be used as fuel in direct alcohol fuel cells, due to its simpler molecular structure and thus easy oxidation. However, methanol is toxic for humans and is mainly produced by fossil fuels. An innovative approach to overcome these disadvantages is to use a mixture of methanol and ethanol, which can be produced by renewable sources, therefore allowing a decrease in the fossil fuels dependency. Additionally ethanol as a larger molecular size, which can be advantageous to decrease the fuel permeation through the membrane from the anode to the cathode side, and has a higher energy density than methanol. Therefore, the goal of this work was to study the effect of using different methanol and ethanol ratios and fuel concentrations on the pDAFC performance towards an increase of its power output and a decrease of the fuel crossover. Additionally, this study was performed using two different MEAs, commercial and tailored MEA, with different diffusion layers on both anode and cathode sides, towards a cost reduction and an enhanced performance. The cell behavior was analyzed through its polarization curves and these results were explained under the light of the EIS data and EEC fitting. The EIS measurements allowed the identification of each loss that negatively affect the cell performance, ohmic losses, anode and cathode activation losses, which included the activation losses due to the oxygen reduction reaction and the activation losses due to the parasitic methanol oxidation reaction at this side. The EEC used in this work showed a good agreement with the experimental data, meaning that reproduces with accuracy the system under study.

The experimental results showed that, for the two MEAs tested, the highest power outputs were obtained using only methanol as fuel, mainly due to the slow electrochemical reaction rates of ethanol on Pt/Ru catalysts. Pt/Ru is widely used as anode catalyst in DAFCs, showing an optimal performance for the methanol oxidation reaction. However, for the ethanol oxidation reaction the use of a binary or ternary catalyst with a higher electrocatalytic activity for this reaction, such as Pt/Sn, would be beneficial. Moreover, despite its higher molecular size, an increase of the ethanol amount on the fuel solution was not beneficial to decrease the fuel crossover rate.

The results also revealed that for lower fuel concentrations similar results were achieved with an ethanol ratio of 25%, 50% and 75% on the fuel solution. These results exposed that if the power requirements are not very demanding it is possible to feed the cell with a lower amount of methanol taking the advantages of ethanol renewability and lower toxicity.

In this work, a maximum power density of 3.00 mW/cm^2 was achieved using the tailored MEA and a methanol concentration of 5 M and 1.58 mW/cm^2 using the commercial MEA and a fuel concentration of 2 M. The tailored MEA besides enabling the operation of the cell with a higher fuel concentration, allowed a reduction of 8% on the system costs and an increase on the power output of 47%.

ACKNOWLEDGMENTS

B.A. Braz acknowledges the Ph.D. fellowship (BEX 12997/13-7) supported by CAPES, "Coordenação de Aperfeiçoamento de Pessoal de Nível Superior" - Brazil. This work was financially supported by: Project PTDC/NewPortCell-POCI-01-0145-FEDER-032116 - funded by FEDER funds through COMPETE2020 - Programa Operacional Competitividade e Internacionalização (POCI) and by national funds (PIDDAC) through FCT/MCTES. POCI (FEDER) also supported this work via CEFT.

REFERENCES

Badwal, S. P. S., Giddey, S., Kulkarni, A., Goel, J. and Basu, S. (2015). Direct Ethanol Fuel Cells for Transport and Stationary Applications - A Comprehensive Review. *Applied Energy,* 145: 80–103. https://doi.org/10.1016/j.apenergy.2015.02.002.

Bambagioni, V., Bianchini, C., Marchionni, A., Filippi, J., Vizza, F., Teddy, J., Serp, P. and Zhiani, M. (2009). Pd and Pt-Ru Anode Electrocatalysts Supported on Multi-Walled Carbon Nanotubes and Their Use in Passive and Active Direct Alcohol Fuel Cells with an Anion-Exchange Membrane (Alcohol = Methanol, Ethanol, Glycerol). *Journal of Power*

Sources, 190: 241–251. https://doi.org/10.1016/j.jpowsour.2009.01.044.

Beyhan, S., Coutanceau, C., Léger, J. M., Napporn, T. W. and Kadirgan F. (2013). Promising Anode Candidates for Direct Ethanol Fuel Cell: Carbon Supported PtSn-Based Trimetallic Catalysts Prepared by Bönnemann Method. *International Journal of Hydrogen Energy,* 38: 6830–6841. https://doi.org/10.1016/j.ijhydene.2013.03.058.

Braz, B. A., Moreira, C. S., Oliveira, V. B., Pinto and A. M. F. R. (2019). Effect of the Current Collector Design on the Performance of a Passive Direct Methanol Fuel Cell. *Electrochimica Acta,* 300: 306–315. https://doi.org/10.1016/j.electacta.2019.01.131.

Carvalho, L. L., Colmati, F. and Tanaka, A. A. (2017). Nickel-Palladium Electrocatalysts for Methanol, Ethanol, and Glycerol Oxidation Reactions. *International Journal of Hydrogen Energy*, 42: 16118–16126. https://doi.org/10.1016/j.ijhydene.2017.05.124.

Gupta, U. K. and Pramanik, H. (2019). Electrooxidation Study of Pure Ethanol/Methanol and Their Mixture for the Application in Direct Alcohol Alkaline Fuel Cells (DAAFCs). *International Journal of Hydrogen Energy*, 44: 421–435. https://doi.org/10.1016/j.ijhydene.2018.03.008.

Herranz, D., Escudero-Cid, R., Montiel, M., Palacio, C., Fatás, E. and Ocón, P. (2018). Poly (Vinyl Alcohol) and Poly (Benzimidazole) Blend Membranes for High Performance Alkaline Direct Ethanol Fuel Cells. *Renewable Energy*, 127: 883–895. https://doi.org/10.1016/j.renene.2018.05.020.

Hu, Q., Li, G., Pan, J., Tan, L., Lu, J. and Zhuang, L. (2013). Alkaline Polymer Electrolyte Fuel Cell with Ni-Based Anode and Co-Based Cathode. *International Journal of Hydrogen Energy,* 38: 16264–16268. https://doi.org/10.1016/j.ijhydene.2013.09.125.

Leo, T. J., Raso, M. A., Navarro, E. and Mora, E. (2013). Long Term Performance Study of a Direct Methanol Fuel Cell Fed with Alcohol Blends. *Energies,* 6: 282–293. https://doi.org/10.3390/en6010282.

Oliveira, V. B., Pereira, J. P. and Pinto, A. M. F. R. (2016). Effect of Anode Diffusion Layer (GDL) on the Performance of a Passive Direct

Methanol Fuel Cell (DMFC). *International Journal of Hydrogen Energy,* 41: 19455–19462. https://doi.org/10.1016/j.ijhydene.2016.06.116.

Oliveira, V. B., Rangel, C. M. and Pinto, A. M. F. R. (2011). One-Dimensional and Non-Isothermal Model for a Passive DMFC. *Journal of Power Sources,* 196: 8973–8982. https://doi.org/10.1016/j.jpowsour.2011.01.094.

Ong, B.C., Kamarudin, S.K. and Basri, S. (2017). Direct Liquid Fuel Cells: A Review. *International Journal of Hydrogen Energy,* 42: 10142–10157. https://doi.org/10.1016/j.ijhydene.2017.01.117.

Park, Y., Kim, D., Lim, S., Kim, S., Peck, D. and Jung, D. (2012). Design of a MEA with Multi-Layer Electrodes for High Concentration Methanol DMFCs. *International Journal of Hydrogen Energy,* 37: 4717–4727. https://doi.org/10.1016/j.ijhydene.2011.04.112.

Pereira, J. P., Falcão, D. S., Oliveira, V. B. and Pinto, A. M. F. R. (2014). Performance of a Passive Direct Ethanol Fuel Cell. *Journal of Power Sources,* 256: 14–19. https://doi.org/10.1016/j.jpowsour.2013.12.036.

Piela, P., Fields, R. and Zelenay, P. (2006). Electrochemical Impedance Spectroscopy for Direct Methanol Fuel Cell Diagnostics. *Journal of the Electrochemical Society,* 153: A1902–A1913. https://doi.org/10.1149/1.2266623.

Song, S. Q., Zhou, W. J., Zhou, Z. H., Jiang, L. H., Sun, G. Q., Xin, Q., Leontidis, V., Kontou, S. and Tsiakaras, P. (2005). Direct Ethanol PEM Fuel Cells: The Case of Platinum Based Anodes. *International Journal of Hydrogen Energy,* 30: 995–1001. https://doi.org/10.1016/j.ijhydene.2004.11.006.

Song, S., Zhou, W., Liang, Z., Cai, R., Sun, G., Xin, Q., Stergiopoulos, V. and Tsiakaras, P. (2005). The Effect of Methanol and Ethanol Cross-over on the Performance of PtRu/C-Based Anode DAFCs. *Applied Catalysis B: Environmental,* 55: 65–72. https://doi.org/10.1016/j.apcatb.2004.05.017.

Wang, J., Wang, H. and Fan, Y. (2018). Techno-Economic Challenges of Fuel Cell Commercialization. *Engineering,* 4: 352–360. https://doi.org/10.1016/j.eng.2018.05.007.

Wongyao, N., Therdthianwong, A. and Therdthianwong, S. (2011). Performance of Direct Alcohol Fuel Cells Fed with Mixed Methanol/Ethanol Solutions. *Energy Conversion and Management,* 52: 2676–2681. https://doi.org/10.1016/j.enconman.2011.01.005.

Xu, C., Faghri, A., Li, X. and Ward, T. (2010). Methanol and Waer Crossover in a Passive Liquid-Feed Direct Methanol Fuel Cell. *International Journal of Hydrogen Energy,* 35: 1769–1777. https://doi.org/10.1016/j.ijhydene.2009.12.055.

Yuan, W., Tang, Y., Yang, X. and Wan, Z. (2012). Toward Using Porous Metal-Fiber Sintered Plate as Anodic Methanol Barrier in a Passive Direct Methanol Fuel Cell. *International Journal of Hydrogen Energy,* 37: 13510–13521. https://doi.org/10.1016/j.ijhydene.2012.06.094.

Zainoodin, A. M., Kamarudin, S. K., Masdar, M. S., Daud, W. R. W., Mohamad, A .B. and Sahari, J. (2014). High Power Direct Methanol Fuel Cell with a Porous Carbon Nanofiber Anode Layer. *Applied Energy,* 113: 946–954. https://doi.org/10.1016/j.apenergy.2013.07.066.

Zakaria, Z., Kamarudin, S. K. and Timmiati, S. N. (2016). Membranes for Direct Ethanol Fuel Cells: An Overview. *Applied Energy,* 163: 334–342. https://doi.org/10.1016/j.apenergy.2015.10.124.

Zakil, F. A., Kamarudin, S. K. and Basri, S. (2016). Modified Nafion Membranes for Direct Alcohol Fuel Cells: An Overview. *Renewable and Sustainable Energy Reviews,* 65: 841–852. https://doi.org/10.1016/j.rser.2016.07.040.

Zhou, W. J., Zhou, B., Li, W. Z., Zhou, Z. H., Song, S. Q., Sun, G. Q., Xin, Q., Douvartzides, S., Goula, M. and Tsiakaras, P. (2004). Performance Comparison of Low-Temperature Direct Alcohol Fuel Cells with Different Anode Catalysts. *Journal of Power Sources,* 126: 16–22. https://doi.org/10.1016/j.jpowsour.2003.08.009.

In: Advances in Energy Research. Volume 31 ISBN: 978-1-53616-271-4
Editor: Morena J. Acosta © 2019 Nova Science Publishers, Inc.

Chapter 5

NOVEL DETERMINATION OF POTENTIAL ENERGY OF COHESION IN SOLIDS AND LIQUIDS: TIME EFFICIENCY OF POTENTIAL ENERGY IN THERMAL PROCESSES

Francisco Ros*
Instituto de Química Médica, CSIC, Madrid, Spain

ABSTRACT

Potential energy in materials relates to the development of interaction energy of electrostatic or electronic character between the particles of the chemical entity constituting a material. The interaction energy develops when the particles are assembled to form the material starting with the particles ideally separated like a gas. The forces of cohesion between atoms, ions or molecules forming materials comprise weak interactions of dipole character, and weak and strong chemical bonding, depending on the type of material. Potential energy of cohesion in solids correlates with the vibrations of the lattice of the solid at the outside of the constituent chemical entity. The dissociation potential energy of the single oscillator

* Corresponding Author's E-mail: iqmr322@iqm.csic.es.

characteristic to lattice vibrations equals the potential energy of the lattice. A new method to find out the cohesion potential energy in solids is described, and fundamental factors internal energy, entropy, temperature and time are examined as connected to potential energy in processes, in particular the connection of entropy and time to potential energy. The method concerns molecular solids and rests upon measurable thermodynamic quantities: heat of sublimation, heat capacity, and entropy as derived from the latter. Valuation of cohesion potential energy by the heat of sublimation requires knowledge of the kinetic energy for molecules both in the solid and the gas, the cohesion energy being the second component of the overall energy in the solid. The vibrational energy for the solid can be known from the heat capacity using instrumentally Debye's theory yet a molecular solid is not monoatomic as prescribed in the theory. The mathematical function for entropy in this theory fits sufficiently better to entropy in the poliatomic molecular solid than the function for heat capacity to the heat capacity of the solid. Accuracy in determining vibrational energy involves the use of the entropy function. Vibrational energy for the gas molecules is difficultly obtainable with sufficient accuracy by heat capacity with simple harmonic oscillator model and can be attained by entropy and Debye entropy function as applied to the gas. Besides cohesion potential energy the method allows for determination of absolute entropies of solids and cognate melts and vapor phases using a restricted set of heat capacity data. In molecular liquids Debye function for entropy is suitable to valuate jointly vibrational and rotational entropies. Cohesion potential energy is embodied into the translation entropy of molecules in a liquid, this feature of potential energy being connected to the relationship of entropy to time. The method yields the cohesion potential energy for a liquid added up to the bringing of internal electronic energy, when this proceeds by facile diaelectrical mobility in the liquid. The efficacy of potential energy in a spontaneous transformation or process with minimal thermic and entropic dispensings relies on the speed of the transition.

Keywords: potential energy of cohesion, solid and liquid states, heat capacity, vibrational entropy, absolute entropies, crystal polymorphism, materials science, irreversibility thermodynamics, entropy-time relationship, maximal entropy

INTRODUCTION

Physical and Thermodynamic Fundamentals on Potential Energy of Cohesion

Potential energy in materials relates to the potential energy connected to cohesive interactions between the chemical particles that bring about a specific material. Being a potential energy as it is, potential energy of cohesion is equivalent to the work required for undoing a material in an imaginary manner by separating particles from one another so that cohesive forces vanish ($E_P = W$, Figure 1). This work is developed against the cohesive forces that tend to preserve the material as such, and has a positive value like the coincident potential energy.

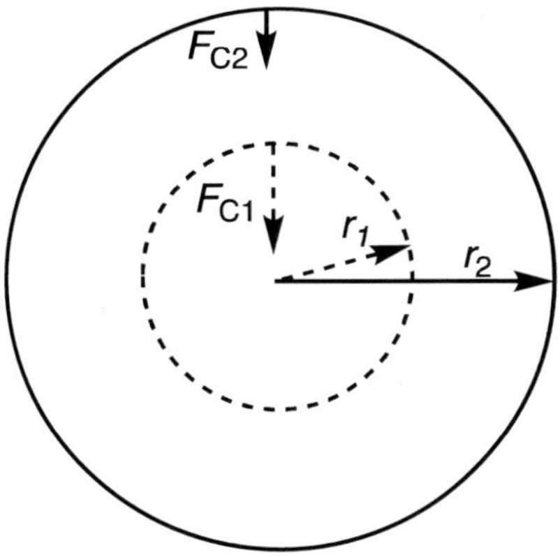

Figure 1. Production of potential energy of cohesion ($dW = Fdr = dE_P$).

Cohesion potential energy actually resides at the uninteracting separated particles, although with regard to the force field of cohesive forces. The potential energy is released, decreasing its value, as the particles assemble to form the material. When a material undergoes a solid-solid or solid-liquid transformation the potential energy changes according to cohesive forces. It thus generally increases on melting or feasibly on a polymorphic transition in the solid state.

Potential energy of cohesion pertains to the content of energy in a material, or the internal energy (equation 1). The kinetic energy ascribable to the motion of particles is the remaining component of the internal energy (equation 2). Internal energy in a perfect state for the removed particles will be only kinetic energy with exclusion of potential energy, doing away with related underlying condensed states promoting positive potential energy. Therefore, a value of zero for the potential energy can be ascribed to such a state. A low-pressure vapor is a real example of a substance whose energy content is just kinetic energy. On occasional condensation the potential energy will manifest as a negative quantity according to its inherent decrease in the transformation, and the internal energy of the condensed material will lessen (equation 3).

$$E_{\mathrm{P}} \equiv U \tag{1}$$

$$U = E_{\mathrm{K}} + E_{\mathrm{P}} \tag{2}$$

$$\Delta U = E_{\mathrm{P2}} - E_{\mathrm{P1}}(0) < 0 \tag{3}$$

In the reverse process, wherein potential energy develops in separating the particles, internal energy increases. For this transformation, first law of thermodynamics, $\Delta U = Q - W$, provides equation 4, disregarding heat at this point. In respect of this law the work implied in the production of potential energy is to be taken as a negative quantity because work is developed by a force external to the cohesion field and pertaining to the material. This internal force acts against the external field and reinforces the internal energy.

$$\Delta U = -W \tag{4}$$

An infinitesimal elemental amount of work in a field of cohesive forces, as produced spontaneously by the cohesive forces, can be expressed in the usual manner for work as $dW_{\mathrm{C}} = F_{\mathrm{C}}dr$, with $dr < 0$ (contrariwise to Figure 1) and $F_{\mathrm{C}} > 0$. Here dW_{C} is negative, but as regards internal energy the effect of the work spontaneously performed by the internal field is to decrease the internal energy because it is a positive work performed by the internal energy. Consequently, force F_C is to be taken as a negative quantity. This also holds for the reverse process wherein potential energy is produced.

Elemental work dW_C can be differently expressed by an internal pressure as a substitute for the cohesive force, and by an elemental change in the bulk volume filled by the separated particles as a substitute for dr (equation 5). Likewise to F_C, the direction of internal pressure p_C is toward contraction. Also, the direction of p_C is the same that an external pressure which will be acting on the particles considered as a bulk thermodynamic phase, but the nature of the latter pressure is mechanical while the former is a result of the existance of an internal cohesion field. In addition, the effect of p_C on internal energy is to diminish the internal energy upon a contraction of volume, contrary to an external pressure and likewise to F_C. The sign of pressure of cohesion p_C as a quantity is minus as it is a pressure which is internal and not external to the system.

$$dW_C = p_C dV \tag{5}$$

The work of cohesion is alike to the external work employed to break apart a solid bar, although in this case cohesive forces operate only in an interfacial direction and not fully around the particles. This surface work of cohesion can be expressed in terms of a surface tension [1] rather than a cohesion pressure, to say $dW_C = \sigma_C dS$. Here cohesion tension σ_C is negative consistently with p_C, and dS is positive as represents a surface enlargement, all which bring about an increase in internal energy (equation 4) consistently with the partial generation of cohesion potential energy.

Forces of Cohesion

Materials are formed of cohesionated atoms, molecules or ions. Cohesive forces [1-3] can be of electrostatic nature as due to static electric charges, as in ionic materials like salts or organic ionic liquids. Electrostatic forces can also result from dipoles, as in molecular solids or liquids whose molecules are polar due to an asymmetric distribution of electric charges in the molecule. Moreover, a dipolar molecule when assembled in a material may induce a polarization of molecules of its kind as a consequence of a nonregular assembling, which results in an alteration of the molecular dipole with respect to the isolated molecule. Dipolar interactions are widely spread in organic materials.

Apart from this, the internal circulatory movement of electrons gives rise in a general manner to transient dipoles that average zero in time, but results in an attractive correlated interaction between neighboring particles, namely London dispersion forces. This kind of weak general interaction is the only cohesive interaction in condensed noble gases and in apolar molecular substances, like tetrachloromethane. London forces become greater with the number of electrons, thus being greater in polymers.

Forces of cohesion can besides be of chemical character as in covalent solids, such as diamond or carborundum, or in metals. In these materials strong bonding between atoms extends through the material in a tridimensional array. Weak directional chemical bonding between molecules, such as hydrogen bonds, is still another kind of cohesive force that is operative in materials.

Conjunction of Potential Energy of Cohesion to Lattice Vibrations

Lattice vibrations refer to oscillatory displacements of the centers of mass of the particles of a material. In monoatomic solids they correspond to the three translational degrees of freedom owned by the isolated atoms in an ideal gas state. In molecular solids the degrees of freedom that would correspond to rotation of the isolated molecule are to be counted for lattice vibrations, which are distinguished from vibrations that are properly internal to the molecule ($3n - 5$, or $3n - 6$ degrees of vibrational freedom for linear or nonlinear molecules with n atoms). Those extra vibrations of the lattice may be regarded as rotations of the molecule as a whole, as if an atom, and which are rotations hindered due to the existence of external cohesion forces pertaining to the lattice. Thus they are ascribable to these external forces rather than to internal forces in the molecule. Lattice vibrations are usually slow oscillations compared with vibrations associated with covalent bonds. The kinetic energy ascribable to a solid is the sum of lattice and molecular vibrations.

Lattice vibrations can be represented in statistical mechanics by a single oscillator with a characteristic average frequency since their number is very large, to say 3*N* vibrations for a monoatomic solid according to the number of atoms in the solid. Hence the potential energy of dissociation of the oscillator representative of lattice vibrations coincides with the potential energy of cohesion of the solid. Thus the lattice vibrations turn out equivalent to the potential energy of cohesion of the solid. The process is sketched in Figure 2. Lattice vibrations are in physical concept totally separable from molecular vibrations in a solid.

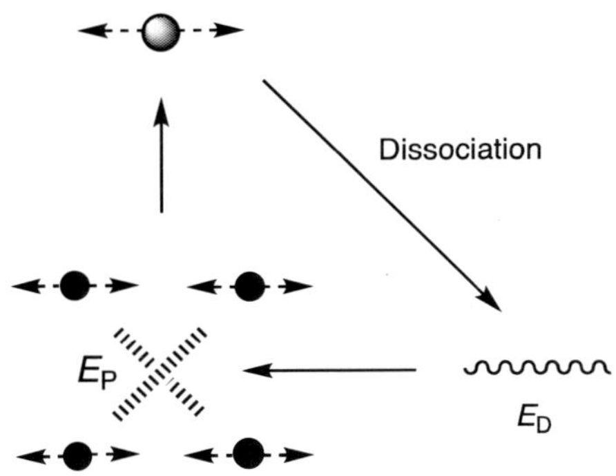

Figure 2. Conversion of lattice vibrations into potential energy of cohesion.

The potential energy of dissociation of the characteristic oscillator for the lattice can be approached by the frequency characteristic to the lattice and by a potential, which gives the potential energy as a function of the binding distance in an oscillator. In Morse potential [4, 5] the dissociation energy for a harmonic oscillator is given as indicated in equation 6 (factor 6 stands for degrees of freedom of lattice vibrations). In monoatomic solids, like metals, covalent materials or simple ionic salts, the average lattice frequency can directly be known by Debye's theory for monoatomic solids [6]. In molecular solids the lattice frequency is to be found out as separated from molecular vibrations.

$$E_{\mathrm{P}} = -6D = -6\frac{2\pi^2 m\bar{\nu}_{\mathrm{L}}{}^2}{\beta^2} \tag{6}$$

Fundamental Methods to Determine Potential Energy of Cohesion

In solids potential energy of cohesion can be obtained by calculation with knowledge of the structure of the lattice and using a potential, which is a mathematical function correlating the potential energy and the distance between two interacting particles [1-4]. In noncrystalline solids the irregularity of the lattice is to be taken into account in the calculations. For molecular solids the potentials must embody the geometry of the molecule for an appropriate description of cohesives forces in respect of the close packing of molecules in the lattice. In addition, the geometry of a molecule (bond lengths and angles) may be it modified in passing to the solid from isolation, as a consequence of polarization effects on proximity. Therefore, the geometry of the isolated molecule is to be considered concerning potentials in order to set a proper thermodynamic reference for potential energy.

In liquids or melts interactions between particles are less directed as a consequence of the translational motion of particles and the rotational motion in molecular materials, all which randomize interactions. Relaxation of cohesive forces as a consequence of these effects is to be taken into account in the potentials for an appropriate description of cohesive interactions.

A different way to obtain the potential energy of cohesion is based on experimental measurement of the energy of sublimation or vaporization of a solid or a liquid instead of using sophisticated theoretical potentials. From the energy for the change of state the potential energy can be found out knowing the kinetic energy for the motion of particles in the gas and condensed states (equation 7), provided that the pressure of the vapor phase is sufficiently low so that potential energy in this phase can be set as a zero reference. The internal energy of change of state is obtained by the experimental heat (Q, equal to the change in enthalpy ΔH at constant pressure) with a correction for mechanical work developed in the phase change ($W = p\Delta V$), which approximately amounts value RT (R gas constant) under frequent conditions.

$$\Delta U = -E_{\mathrm{P}} + \Delta E_{\mathrm{K}} = Q - W \tag{7}$$

Measurement of the heat of sublimation or vaporization is usually carried out by the change in vapor pressure with temperature and applying Clausius-Clapeyron equation to the system (equation 8) [7]. A minimal detectable vapor pressure of the material at the temperature of experiment is necessary to get in this way the energy of transformation to the gas state.

$$Q = RT^2 \frac{d\ln p}{dT} \tag{8}$$

When the chemical species in gas state is not the same that in condensed state the method is to be used in combination with a thermochemical cycle. The method can be also used to obtain the change in potential energy in solid-solid or solid-liquid transformations using the corresponding heats. An example of the method for these transformations is described in the chapter together with solid-vapor and liquid-vapor transformations.

The kinetic energy of a solid, which is the energy of lattice and molecular vibrations, can be found by the heat capacity of the solid ($C = dU/dT$) using theories of statistical thermodynamics for heat capacity of solids based on the quantic harmonic oscillator [6, 8, 9]. In liquids and gases the translational and rotational energies of molecules add to molecular vibrations for the total kinetic energy, each of the former energies being equal to quantity $3RT/2$. It will be shown in the chapter that the remaining vibrational energy in liquids and gases can be obtained by the heat capacity using theories for heat capacities of solids.

The strength of the vibrations of an isolated molecule is relaxed in passing to a condensed state as a consequence of the dielectric nature of matter. The vibrational frequency and concomitant force constant of an oscillating electric dipole decreases in a dielectric medium [10]. This alteration of force constant conveys an increase in the electronic potential energy of binding of the oscillator (as if a molecule expanded in condensation from the gas state) and it adds to the decrease in potential energy of cohesion on condensation.

This merge of potentials energies of different character (internal or external to the molecule) and sign (the former increasing, the latter decreasing) is to be evaluated in obtaining the potential energy of cohesion by this method. The energetic effect of the change in force constant may not be large as according to Born-Oppenheimer approximation the electronic motion is faster than the vibrational nuclear motion. Thus the effect of a vibration on the electronic binding energy of the oscillator is relieved. On the other hand, potential energy of cohesion in molecular substances is small in comparison with the electronic binding energy and the former may be hidden in a small change in electronic potential energy. The effect may be larger in liquids due to a higher dielectrical mobility in the liquid medium which enhances electronic energy.

Special Features of Potential Energy Concerning Entropy and Time

Potential energy in materials itself lacks entropy. It follows Planck's special enunciation of third law of thermodynamics according to which entropy in internally perfectly ordered materials will be null when the kinetic energy of particles and of thermal origin vanishes at absolute zero temperature, whereas potential energy actually persists in a material in such an event of null entropy. Thermodynamic entropy is a property of matter filling up a bulk space wherefrom cannot be separated, and generally depends on temperature. Neither attribute is possessed by potential energy. However, potential energy affects entropy as is incorporated into internal energy ($U = E_K + E_P$) whereby is connected to entropy, to be specific $\Delta U = Q - W$ and $dQ = TdS$.

Thermodynamic entropy can be of either thermal or spacial character. The latter results from particles going to a space by a thermic displacement, and is independent of temperature. In an expansion of a perfect gas at constant temperature, thermodynamic entropy increases while the energy of motion of molecules as well as their thermal entropy remains constant. Such an increase in entropy is merely due to athermal spacial entropy, that is $\Delta S = R\ln(V_2/V_1)$ [7]. It should be remarked at this point that thermodynamic spacial entropy is prior to statistical considerations ($k_B = R/N_A$).

According to the precedent thermodynamic formula, spacial entropy actually occurs in the absence of potential energy of interaction between molecules. Therefore, potential energy is not needed for spacial entropy. The perfect gas consists of microscopic particles undergoing an entirely free displacement, which is a motion of thermic nature and within the bulk volume of the gas. Thus thermodynamic spacial entropy is undissociable from the presence of finite matter and free displacement of particles [$\Delta S = R\ln(V_2/V_1)$]. Empty space, space itself, lacks entropy.

In nonperfect gases potential energy brings about a confinement of spacial entropy together with a lowering of the overall translational entropy. This is discussed below in connection with the translational entropy of liquids. Mixing and configurational entropies are another kinds of entropy that do not depend on temperature, and in essence are forms of spacial entropy as a diversity of material entities come into a room.

With regard to the above perfect gas, a loosening of cohesive interactions on expansion of a solid material weakens lattice vibrations. The effect is to diminish the underlying zero-point energy for these vibrations and to enhance their uprising thermal energy as well as their entropy. In this manner, thermal entropy increases in an isothermal expansion of a solid whereas remains constant in such an expansion of a perfect gas. As concerns spacial entropy, such effects on expansion of a solid can be considered as taking place on the single oscillator representing lattice vibrations. Thus it has not to do with spacial entropy as an oscillation takes place just about a point. Vibrations do not contribute to spacial entropy. Regarding internal molecular vibrations in a molecular solid the effect of a loosening of cohesive forces should be to relax the vibrations, but this is counterpoised by the dielectric effect that operates in the opposite way as it has been above indicated in the preceding subsection. These effects on isothermal expansion of condensed materials are small.

Entropy in Liquids

The entropy of liquids serves to clarify preceding concepts and permits an insight to the role of potential energy in entropy. Only translational entropy in a liquid is here considered owing to its particular significance to potential energy of cohesion, and according to the way that translational entropy of molecules in a liquid is separable from the vibrational and rotational entropies. This abridged stage amounts to a monoatomic liquid. Entropy does not do away with itself in a transformation because is a fixe quantity of state of a body, likewise internal energy is, but converts in a balanced manner, and can be created or destroyed in a transformation. Entropy can also transform into different forms of entropy.

For liquefaction of the considered liquid from the gas state the balance of entropy can be set as equation 9. When the liquefaction takes place at constant temperature and pressure the change in entropy (ΔS) equals Q/T, which in turn is expressed by the right hand side of equation 10, in accordance with what has been above indicated in a subsection (Fundamentals methods to determine potential energy of cohesion). The kinetic energies for the liquid and the gas are $3RT/2$ and annul each other in the equation. Term E_P refers to the potential energy of cohesion of the liquid, as the gas phase is taken as a perfect gas having zero potential energy (there exists no interaction between particles). Then using Sackur-Tetrode equation for the entropy of a perfect monoatomic gas [7] equation 11 straightforward ensues for the entropy of the monoatomic liquid.

$$S_L = S_G + \Delta S \tag{9}$$

$$\Delta S = \frac{\Delta U}{T} + \frac{p\Delta V}{T} = \frac{E_P}{T} + \frac{p\Delta V}{T} \tag{10}$$

$$S_L = R\left[\frac{3}{2} + \ln\left(\frac{2\pi mkT}{h^2}\right)^{3/2}\right] + R\left(\frac{pV_L}{RT} + \ln V_G\right) + \frac{E_P}{T} - R\ln N_A \tag{11}$$

$$S_L = R\left[\frac{3}{2} + \ln\left(\frac{2\pi mkT}{h^2}\right)^{3/2}\right] + R\left(\frac{pV_L + E_P}{RT} + \ln\frac{V_G}{N_A}\right) \tag{11´}$$

The first term in equation 11 stands for the thermal entropy of translational motion. Adimensional factor 3/2 relates to the actual energy for the three degrees of translational motion, and the accompanying logarithmic quantity elicits the relationship of thermal entropy to temperature. The second term in the equation formally refers to the spacial entropy concerning the translational motion. With analogy to 3/2, adimensional pV_L/RT relates to the actual space filled by the liquid, while the corresponding logarithmic term refers to the collateral space of reference.

The last two terms in equation 11, which comprise the potential energy, are formally analogous to the preceding terms in the equation from algebraic viewpoint. It thus appears that they conform the entropy of potential energy, similarly to the preceding terms concerning thermal and spacial entropies. However, the particular term of reference $R\ln N_A$ is void of a physical meaning other than statistical, which results in a potential-energy entropy of merely statistical character, not entirely thermodynamic and not consubstantial with potential energy. Moreover, such a sort of entropy is a negative entropy since term E_P/T is a negative quantity as above pointed out, and term $R\ln N_A$ is negative. A negative value for an absolute entropy is not sound.

Potential energy at a referential zero value may positively exist in space, and when is released the actual spacial entropy simultaneously reduces according to equation 11. According to third law, potential energy itself is deprived of entropy, as above mentioned, and if potential energy ought to give rise to potential-energy entropy concerning liquids or nonperfect gases it would be a casuistic case.

In an alternative and more reasonable way, equation 11 can be rearranged so that term E_P/T is correlated to the volume of the liquid and hence to spacial entropy (equation 11´). Term $R\ln N_A$ from original Sackur-Tetrode equation is restored by this operation to its statistical role in spacial entropy instead of being an attribute of dubious potential-energy entropy. In the manner of equation 11´ the meaning of potential energy to spacial entropy, being E_P a negative quantity, is to reduce the space entropically available to translational motion, what is physically meaningful (Figure 3). In this way a potential-energy entropy is assimilated into real and genuine spacial entropy (provided molecular translational motion exists) and thereby renouncing physical identity.

This outcome is logical since potential energy is of spacial nature, and supports a nonentropic nature of potential energy itself. The operative potential-energy term in equation 11´ is adimensional E_P/RT and not entropy E_P/T as it could be. This points out that the role of potential energy in entropy is merely instrumental. Nevertheless, adimensional factor E_P/RT is actually and fully an entropic factor in the way that results from equation 11´.

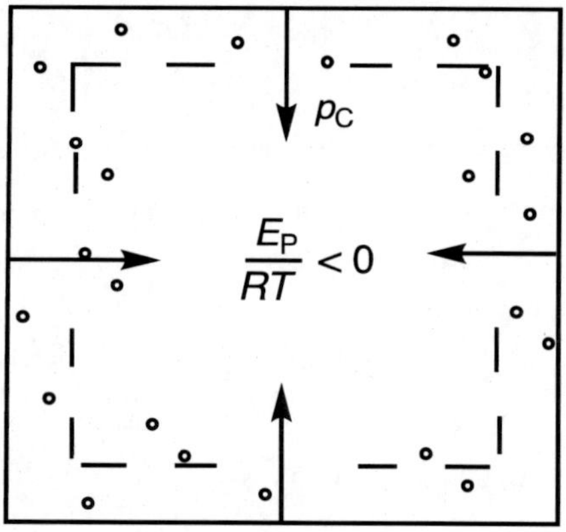

Figure 3. Contraction of entropy space by internal pressure of potential energy of cohesion.

Time Feature

Together with energy, entropy, and temperature, time is the fourth physical magnitude of fundamental importance in thermodynamics. The relevance of time derives from second law of thermodynamics establishing that in a thermally isolated system ($Q = 0$) entropy increases in a process.

The relation of increased entropy to time ensues the fact that natural transformations take a definite time to occur. The general equation correlating entropy and time for a thermally isolated system not producing work is disclosed in the chapter.

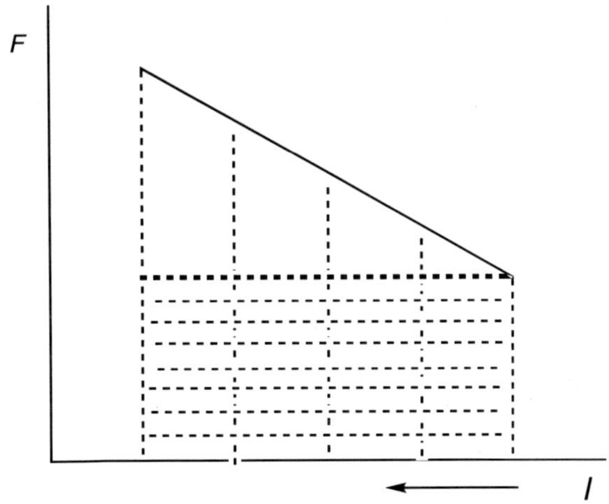

Figure 4. Work in an irreversible process ($W = F\Delta l$).

Second law requires that, for the increase in entropy, a process is to be irreversible, that is to say stages which are not internally equilibrated with respect to a fixe equation of state. As regards potential energy, an irreversible process is sketched in Figure 4 in which F represents a force in a potential energy field (similarly to Figure 1), the slanted line the reversible process and the horizontal line the corresponding irreversible process.

The excess work in the reversible path with respect to the irreversible path is transformed into heat (no useful work) in the latter process. In the case of a thermally isolated system ($Q = 0$) the heat involved in an irreversible process is not the usual external heat being transferred between bodies, but an *internal* heat directly connected to entropy ($Q = T\Delta S$). This is in conformity with second law, to say internal irreversible $Q > 0$ and $\Delta S > 0$. It will be shown in the chapter that irreversibility implies a contraction of time for a process be completed, and this involves potential energy in materials.

METHOD TO DETERMINE THE POTENTIAL ENERGY OF COHESION OF AN ORGANIC COMPOUND IN SOLID AND LIQUID STATES

The potential energy of cohesion of the ester shown in Figure 5, as both a solid and a liquid, has been investigated by the method of the heat of phase change and the kinetic energies for the phases (equation 7, 8). The ester is chiral as possesses an asymmetric carbon atom, and thus exists as two distinct enantiomers, which are not geometrically identical since are not superposable (just mirror images) and are materially separable [11].

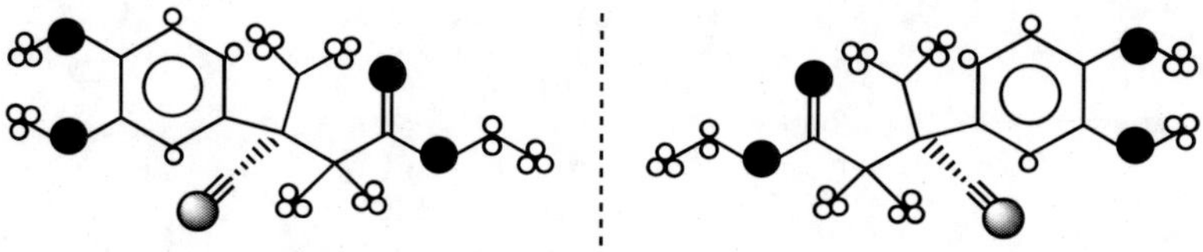

Figure 5. Chiral ester

In the solid state the potential energy has been investigated on the ester as a racemic conglomerate of the solid separate enantiomers (in racemic equal amounts) and as a racemic solid mixture of the enantiomers. In the latter, opposite enantiomers are present in a single crystal lattice [12, 13]. The potential energy of cohesion for the two different enantiomeric crystals existing in the conglomerate is the same (as the crystals are mirror images of each other) and should differ from the potential energy for the crystalline solid mixture in which the two enantiomers are present in a crystal.

In the liquid state the potential energy of cohesion has been investigated on the liquid racemate of the ester. In similarity to the conglomerate with respect to the solid mixture, potential energy for the liquid racemate will not be the same that for the separate liquid enantiomers provided that the racemate is not a perfect solution of enantiomers. This event corresponds to a null heat of solution and an ideal entropy of mixing with value $\Delta S = R\ln 2$. Deviation from a perfect solution of enantiomers should be however very small [12, 14].

Solid Ester

As the material used for the research, the ester was a crystalline racemic conglomerate melting at 91°C, as determined by differential scanning calorimetry (DSC [15]). The two separate enantiomers present in the racemic conglomerate independently melt at 112-114°C, and their conglomerate mixture with minimal melting point corresponds to the racemic conglomerate melting at 91°C (an eutectic mixture [12]). For finding out the heat of sublimation the vapor pressure of the conglomerate was determined by Knudsen method of molecular effusion [1, 16] in a temperature range below the melting point (Table 1). The vapor pressure of either solid enantiomer in the conglomerate is the same. The heat of sublimation at average temperature 79°C was derived from the data using equation (8), and subtraction of small quantity *RT* for work developed in sublimation gave the internal energy of sublimation, 120 kJ mol^{-1} (29 kcal mol^{-1}).

As a first step to obtain the kinetic energy for the solid ester the heat capacity of the solid as a function of temperature was determined by DSC [17] in a temperature range, furnishing adjusted equation 12 (*C* in J K^{-1} mol^{-1} units). This equation refers to heat capacity at constant pressure according to the experimental conditions.

Table 1. Vapor pressure of the solid ester as a conglomerate (1 Pa ≡ 0.01 mbar)

Temperature (°C)	Vapor pressure (Pa)
71	0.0155
77	0.0326
80	0.0474
83	0.0700
86	0.0985

$$C = -1.02 \times 10^{-4}T^2 + 2.39T - 217 \quad (12)$$

The theories correlating heat capacity of a solid with vibrational energy require that heat capacity is at constant volume and not at constant pressure so that vibrational energy correlates with internal energy as it is without mixing with work ($C_V = dU/dT$, $C_p = dQ/dT$). In a molecular solid such as the ester, heat capacity at constant volume may include a trace of potential energy of cohesion as a consequence of a slight variation of the potential energy with temperature, which would arise from a variation of lattice vibrations with temperature due to a nonspherical configuration of the molecule. So potential energy of cohesion of a molecular solid will not be independent of temperature at constant volume, but the effect will be small.

Experimental determination of heat capacity at constant volume is more involved than at constant pressure, so equation 12 was used for a general and preliminary approach to the vibrational energy. The value of potential energy obtained in the end in this way was compared with that subsequently obtained using a refined methodology circumventing heat capacity.

For obtaining the vibrational energy from the heat capacity, Debye's theory [6] was choosen in preference to Einstein's or Nernst-Lindemann's theory [8, 9]. Debye's theory is more realistic as considers a broad range of vibrational frequencies in a solid, while the latter theories assume a single or dual frequency. This is particularly relevant to the present case of a molecular solid which has a diversity of internal molecular vibrations, contrasted with monoatomic solids or simple ionic salts to which all those theories originally refer. Debye function provides heat capacity as a function of a constant temperature characteristic of the particular solid and of the temperature of the solid, that is $C = Nf(T_D/T)$. Value N refers to the number of atoms in a molecule (presently $N = 51$) so that all vibrations in the solid are taken into account including lattice vibrations ($3N = 153$ total degrees of freedom for the solid ester).

Adequacy of Debye function can be tested by the constancy of T_D in applying experimental C-T values on the function. The T_D value corresponding to each C-T pair can be solved by the tabulated function [18]. Using equation 12 in this manner for C-T values in a broad temperature range, the corresponding T_D values agreed within 0.4%, $T_D = 1456 \pm 6$ K. With this T_D value the kinetic energy was found applying the value on the

Debye function for vibrational energy, $E = Nf(T_D/T)$, which is in strict correspondence with the heat-capacity function [18, 19]. Debye energy function provides directly the thermal vibrational energy, excluding zero-point energy. This is added to the former to obtain the total vibrational energy and amounts to $E_0 = 9NRT_D/8$ [1]. In this way the total kinetic energy of the solid ester was 771 kJ mol^{-1} at 79°C sublimation temperature, of which a 90% corresponds to E_0 due to strong vibrations in the covalently bound molecule.

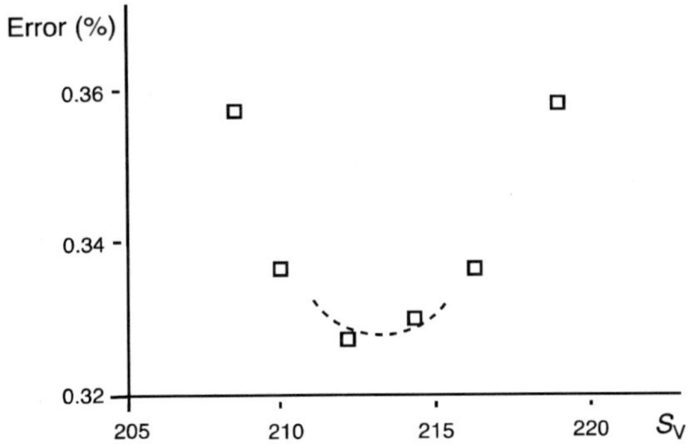

Figure 6. Maximal fitting to entropy for the ester in gas state (S_V in J K^{-1} mol^{-1}).

The vibrational energy for the ester in the gas state was first approached by heat capacity similarly to the solid ester (translational and rotational energies for the isolated molecule are $3RT/2$, and the respective heat capacities $3R/2$). The heat capacity of the gaseous ester was not experimentally available and was estimated using Rihani and Doraiswamy's method [20] (equation 13, perfect gas state consistently with the low vapor pressure of the ester). The heat capacity corresponding to vibrational motion alone was obtained subtracting quantity $4R$ concerning the conversion to constant volume from constant pressure (R) as well as the translational and rotational motions ($C = 3R/2$ each). Using equation (13) in combination with the simple heat-capacity function for the quantic harmonic oscillator (Planck-Einstein oscillator) [8, 18], which is a logical choice for the isolate molecule of the gaseous ester, gave a characteristic temperature with an error of 5%, proceeding to the test in the manner for the solid ester. Debye heat-capacity function, which refers to the solid state rather than to the gas state, gave a similar error. Such errors are too large as tenfold compared with that for the ester in the solid state.

$$C = -0.001T^2 + 1.74T - 26.9 \tag{13}$$

The entropy of the gaseous ester was then tried as an alternative to heat capacity. Equation 14 gives the vibrational entropy of the ester in the gas state and was derived from equation 13 by integration between limits ($dS = CdT/T$). Setting an approximate estimated value to indeterminate reference $S_V(79°C)$ and using the equation in combination with Debye entropy function—which is in strict correspondence with the heat-capacity function yet excluding translation and rotation for the isolate molecule, $S = (N - 2)f(T_D/T)$—it was tested in the above manner for heat capacity that the entropy function fitted to the entropy equation better than the heat-capacity function to heat capacity. Then the tentative reference value of entropy in the entropy equation was varied to reach a maximal fitting of the function to the equation, determining in the usual manner the error in T_D for each trial entropy value (Figure 6). This optimal point was taken as the correct entropy value, $S_V = 213$ J K^{-1} mol^{-1}, $T_D = 1706$ K. Finally this T_D value was applied to Debye energy function to obtain the thermal vibrational energy at the sublimation temperature, and the corresponding zero-point energy was added. The total kinetic energy of the ester in the gas state, which is equal to the total internal energy, was 844 kJ mol^{-1} (202 kcal mol^{-1}, including minor energy $3RT$ for translational and rotational energy).

$$S_V = -5 \times 10^{-4}T^2 + 1.74T - 60.2\ln T - 198 + S_V(79°C) \tag{14}$$

From the data obtained and using equation 7 the potential energy of cohesion of the solid ester was −47 kJ mol^{-1} (−11.2 kcal mol^{-1}). This value is consistent with the potential energies of cohesion obtained by methods of exclusive theoretical nature for other molecular compounds [1]. It contrasts with the much higher cohesion potential energy in metals and salts. The total internal energy is 724 kJ mol^{-1} (173 kcal mol^{-1}, $U = E_K + E_P$). Such an absolute content of energy refers to a solid as a chemical substance, with independence of nuclear dismutations, and is the reference energy to chemical transformations or changes of state of a solid, as in the solid-gas transformation considered. Examples for solid-solid and solid-liquid transformations will be provided below.

The novelty of the method resides in the adequacy of Debye functions (which originally refer to just monoatomic solids) to determine the vibrational energy of a molecular solid or gas by means of the heat-capacity or entropy function. In addition the adequacy of the entropy function with respect to the heat-capacity function turns out improved. As regards an isolate molecule the method provides a simple approach to vibrational energy, which is an alternative to knowledge of all frequencies for the normal modes of vibration of a molecule and to the use of an energy function for the oscillations [21].

Refined Method

Accuracy in determining vibrational energy in this method is critical since this energy is much larger than potential energy of cohesion, and a small absolute error in the former may prevent a correct knowledge of the latter as obtained by equation 7. Thus a deviation of 10 kJ mol^{-1} in the vibrational energy of the solid ester (a 1% error) represents a 20% of the potential energy, according to above data. Beside accuracy there exist the unsurenesses about the experimental heat capacity of solid ester, which is at constant pressure and not at constant volume as required by the theory, and about the heat capacity of the gaseous ester which is more or less a coarse estimation. Consequently, an alternative procedure was devised which provides vibrational energies with high accuracy and eludes heat capacities.

The starting equation is the balance of entropy for sublimation with reference to the translational entropy of the gaseous ester (equation 15, S_R and S_V rotational and vibrational entropies of the gas]. The entropy of the solid ester is only vibrational entropy since the ester here refers to one single component and is a regular ordered crystal in addition. Moreover, the entropy of potential energy of cohesion is null according to third law of thermodynamics. The singularity of this procedure resting on entropy is that it cannot be carried out on internal energy because potential energy is an extra unknown quantity additional to vibrational energy, whereas potential energy does not play a role in the entropy of the solid ester which lacks of potential-energy entropy, as pointed out in Introduction (Special features of potential energy concerning entropy and time).

$$S_{\mathrm{T}} = \Delta S_{\mathrm{S}} + S(\mathfrak{s}) - S_{\mathrm{R}} - S_{\mathrm{V}} \qquad (15)$$

The actual translational entropy of the gaseous ester is compared in the procedure with that one predicted by equation 15 using trial values for the entropy of the solid [$S(\mathfrak{s})$], and the value that correctly predicts the translational entropy is taken as the entropy of the solid. Actual S_{T} can exactly be known from experimental vapor pressure and temperature data with Sackur-Tetrode formula for the translational entropy of a perfect gas [$S_{\mathrm{T}} = f(p,T)$] [7]. For this purpose the data in Table 1 as correlated by equation $\ln p = 38.7187 - 14760T^{-1}$ (in Pa) was used, thus obtaining a set of experimental translational entropy values of reference for the test. The test was conducted at different temperatures in the range in Table 1, in order to keep in the experimental interval without extrapolation for reason of needed precision. Numerical values were likewise handled with four or five digits in the test.

The experimental entropy of sublimation amounts to 348.8 J K^{-1} mol^{-1}, as obtained from data in Table 1 passing through the heat of sublimation ($\Delta S = \Delta H/T$), and refers to 351.8 K (78.6°C) as middle temperature. For knowing the ΔS_{S} value at other temperatures in the experimental interval, general Planck equation for the change in heat of change of state with temperature [22] can be handled. For sublimation, this equation can be expressed in a shortened manner as $d\Delta H_{\mathrm{S}} = [C(\mathrm{g}) - C(\mathfrak{s})]dT$, as a very good approximation enough for the present precision test. Integration of this differential equation between limits gives equation 16, which divided by temperature provides the variation of entropy of sublimation for the temperature interval around the middle temperature, as required for the test (equation 17).

$$\Delta H_{\mathrm{S}} = \int_{351.8}^{T} [C(\mathrm{g}) - C(\mathfrak{s})]dT + \Delta H_{\mathrm{S}}(351.8\ \mathrm{K}) \quad (16)$$

$$\Delta S_{\mathrm{S}} = \frac{1}{T}\int_{351.8}^{T} [C(\mathrm{g}) - C(\mathfrak{s})]dT + \frac{\Delta H_{\mathrm{S}}(351.8\ \mathrm{K})}{T} \quad (17)$$

The remaining terms S_{R} and S_{V} in starting equation 15 are expressed by equation 18 and 19 as functions of temperature ($C_{\mathrm{R}} = 3R/2$). As a matter of fact, these latter equations are mathematically and fully embodied in the preceding equation 17 even though algebraic forms are different. Therefore, equation 18 and 19 were offset in 17 in a mathematically exact manner in order to avoid the use of a dubious heat capacity for the gaseous ester and to simplify the test. This operation resulted in equation 20 with its term $\Delta S_{\mathrm{S}}^{\dagger}$ as expressed by equation 21. This equation is simpler than starting 17 and is definite for the test. It permits prediction of the translational entropy of the gaseous ester by equation 20 for a range of temperatures and trial values for the entropy of the solid ester, making use only of the experimental heat capacity of the solid ester and one experimental value of translational entropy.

$$S_{\mathrm{R}} = \int_{351.8}^{T} \frac{C_{\mathrm{R}}}{T} dT + S_{\mathrm{R}}(351.8\ \mathrm{K}) = \frac{3}{2} R\ln\frac{T}{351.8} + S_{\mathrm{R}}(351.8\ \mathrm{K}) \quad (18)$$

$$S_{\mathrm{V}} = \int_{351.8}^{T} \frac{C_{\mathrm{V}}}{T} dT + S_{\mathrm{R}}(351.8\ \mathrm{K}) \quad (19)$$

$$S_{\mathrm{T}} = S(\mathfrak{s}) + \Delta S_{\mathrm{S}}^{\dagger} \quad (20)$$

$$\Delta S_{\mathrm{S}}^{\dagger} = \frac{1}{T}\left\{\int_{351.8}^{T} \left[\frac{5}{2}R - C(\mathfrak{s})\right] dT\right\} + \frac{351.8[S_{\mathrm{T}}(351.8\ \mathrm{K}) - S(\mathfrak{s})(351.8\ \mathrm{K})]}{T} \quad (21)$$

Different trial Debye temperatures for the solid ester in a wide interval were applied to equation 20 and 21. For each T_{D} value a number of values of $S(\mathfrak{s})$ were used in the selected temperature interval [$S = f(T_D/T)$]. Table 2 collects the data for a test Debye temperature. The error in prediction for a particular T_D was assessed by the overall deviation of the calculated S_T set from the reference set previously obtained as above described. Figure 7 shows the optimal fit having perfect coincidence with experiment and which corresponded to 1391 K. By this T_D value the kinetic energy of the solid ester is 747 kJ mol^{-1} (79°C), applying the T_D value to Debye energy function [$E = Nf(T_D/T)$] plus zero-point energy. The kinetic energy above obtained by the unrefined method deviates a 3% (24 kJ mol^{-1}) from this refined value.

Table 2. Sample assay for the refined method (T_D = 1376 K). Entropies in J K^{-1} mol^{-1}

Temperature (K)	$S(\mathfrak{s})$	$\Delta S_S^\dagger$	S_T (predicted)	S_T (actual)
344.3	332.26	−27.11	305.15	314.56
347.3	338.02	−31.86	306.16	311.67
350.3	343.61	−36.58	307.03	308.82
353.3	349.08	−41.29	307.79	306.02
356.3	356.81	−47.98	308.83	303.27
359.3	360.37	−50.64	309.73	300.57

By the T_D value obtained the entropy of the solid ester was 339 J K^{-1} mol^{-1} (79°C), as it followed application of the T_D value to Debye entropy function. The total entropy of the solid ester is just vibrational as indicated in Introduction, so this value amounts to an absolute entropy. The corresponding total absolute entropy of the gaseous ester was 688 J K^{-1} mol^{-1} (79°C), using the experimental entropy of sublimation [$S(g) = S(\mathfrak{s}) + \Delta S_S$].

Confirmation of the vibrational entropy for the gaseous ester, a value of which was obtained by the unrefined method, was required concerning the kinetic energy in the gas state. For this purpose, equation 15 was used in respect of vibrational entropy and for which the rotational entropy of the gaseous ester was needed. This was calculated with the formula for the rotational entropy of the molecule [10]. This formula relates to the product of the three moments of inertia of the molecule, which was calculated by exact Hirshfelder-Wilson method [23, 24] with the coordinates of the atoms in the molecule, which were available by X-ray crystallography [25] ($I_A I_B I_C = 2.521 \times 10^{-131}$ kg^3 m^6).

Figure 7. Maximal fitting in the refined method (deviation is difference of predicted to experimental vapor pressure at 79°C, in Pa).

A moment of inertia regarding the molecule in a crystal will not be exactly the same as in gas state because bond distances and angles in the molecule will be modified from a state to the other as a consequence of cohesion forces and, in addition, molecular conformers with spacially disparate distribution of mass exist in the gas state. Nevertheless, the effect of these alterations on rotational entropy is negligible, and the $I_A I_B I_C$ value obtained with X-ray data was safely used in the formula for the rotational entropy. This amounted to 148 J K^{-1} mol^{-1} (79°C), which in turn gave a highly reliable vibrational entropy of 233 J K^{-1} mol^{-1} (79°C) by equation 15. This vibrational entropy of the gaseous ester corresponded to a T_D value of 1642 K.

By this T_D value the total kinetic energy of the gaseous ester was 820 kJ mol^{-1} [196 kcal mol^{-1}, 79°C, minor energy $3RT$ added to vibrational energy $E = (N - 2)f(T_D/T)$ plus E_0], which amounts to the internal energy (no potential energy). The corresponding value obtained by the unrefined method deviates a 3% (24 kJ mol^{-1}) from this refined value. Using the refined values a potential energy of cohesion of –47 kJ mol^{-1} (–11.2 kcal mol^{-1}) resulted by equation 7. It coincides with the value obtained by the unrefined method because otherwise significative deviations in the unrefined method are counterbalanced in equation 7. The refined total internal energy of the solid ester amounts to 700 kJ mol^{-1} (167 kcal mol^{-1}, $U = E_K + E_P$) versus unrefined 724 kJ mol^{-1}.

As shown, the method affords determination of absolute entropies of a solid and a gas using a restricted set of heat capacity data, which is in contrast with the comprehensive heat capacity range in the classical method for absolute entropies [7]. The absolute entropy of a liquid can be obtained from that for a solid with enlarged heat capacity data, as it will be below shown concerning the liquid ester.

Polymorphic Ester

The potential energy of cohesion of a racemic crystalline polymorph of the solid ester was investigated. The polymorph melted at 83°C as determined by DSC. In this material the two enantiomers of the ester coexist in a single crystal lattice in contrast with the ester as a conglomerate. The internal energy of the polymorph is that of the conglomerate plus the internal energy of formation of the former from the latter (equation 22, A conglomerate, B polymorph).

$$U_B = U_A + \Delta_F U_B \tag{22}$$

Such a formation of the polymorph is a nonchemical transformation in the solid state wherein the crystal lattices of the separate enantiomers become juxtaposed in a single crystal lattice. The temperature selected for determining this heat of formation was the melting point of the polymorph for experimental convenience. At melting point the polymorph and its liquid melt coexist, and the melt is the same liquid that would be obtained undercooling a melt of the conglomerate, whose melting point (91°C) is higher than that of the polymorph.

For knowing the term $\Delta_F U_B$ the heats of melting of the conglomerate and the polymorph were required and were determined by DSC. Heat capacity of the liquid melted polymorph, which was also necessary, was determined by DSC above melting point (equation 23, J K^{-1} mol^{-1}, at constant pressure). This was a complement to heat capacity of the solid conglomerate previously determined (equation 12) that was here used as well. Using the experimental data the term $\Delta_F U_B$ and the final internal energy of solid polymorph were obtained by the summation process shown in Figure 8. Values labeled as Q in the figure refer to constant pressure according to the experimental data ($Q = \Delta H$) and are in practice identical to the corresponding internal energies from which differ in a negligible amount ($< 10^{-3}$ kJ mol^{-1}).

$$C = -0.0112T^2 + 10.2T - 1550 \tag{23}$$

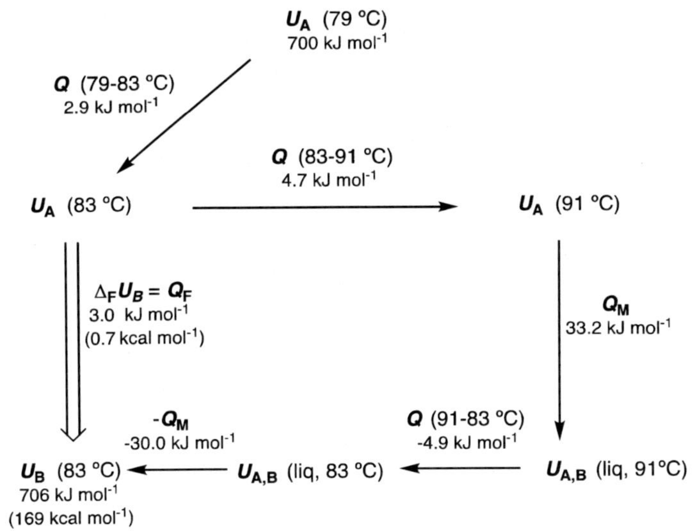

Figure 8. Internal energy of the polymorphic ester (U_B; A conglomerate ester).

For finding out the vibrational energy of the polymorph (as kinetic energy in U_B) the entropy was obtained, as the first step, from the entropy of the conglomerate and the entropy of formation of the polymorph (equation 24).

$$S_B = S_A + \Delta_F S_B \tag{24}$$

The entropy of formation of the polymorph was obtained similarly to the internal energy of formation in Figure 8 thus using equations for entropy as a function of temperature for the conglomerate and liquid melt (equation 25 and 26 derived from experimental 12 and 23, $dS = CdT/T$), and entropies of melting (from Figure 8, $S_M = Q_M/T_M$). The resulting entropy of formation was 6.4 J K^{-1} mol^{-1} (83°C). This value is practically the same as the general entropy of mixing in 1:1 ratio ($\Delta S = R\ln 2 = 5.8$ J K^{-1} mol^{-1}) what indicates that the polymorph is a racemic solution of enantiomers in the solid state rather than a mixed crystal (or a so-called racemic compound [12]). In the solid solution the enantiomers will appear superposed in crystallographic X-ray analysis.

$$S = -5.09 \times 10^{-5}T^2 + 2.39T - 217\ln T \tag{25}$$

$$S = -5.6 \times 10^{-3}T^2 + 10.2T - 1550\ln T \tag{26}$$

The S_A value for the conglomerate at 83°C, as required for equation 24, was obtained by the refined characteristic Debye temperature of the conglomerate (the solid ester) directly using Debye entropy function. This resulted in 348 J K^{-1} mol^{-1}. The added-up entropy of the polymorph (S_B) was 354 J K^{-1} mol^{-1}. Athermal entropy of mixing ($\Delta S = R\ln 2$) is not to be considered in using Debye functions that refer only to thermal quantities. Thus quantity $R\ln 2$ was subtracted from S_B assuming the polymorph be a solid solution as above pointed out, and the neat value was used to obtain the vibrational energy by Debye entropy and energy functions in the usual manner. This resulted in a vibrational energy of 750 kJ mol^{-1} (83°C), which together with the value of total internal energy in Figure 8 (U_B) gives a potential energy of cohesion of –44 kJ mol^{-1} (−10.5 kcal mol^{-1}, $E_P = U - E_K$). This value is very close to that for the solid ester (–47 kJ mol^{-1}), what is logical as enantiomers are so similar that it does not make much difference having either in neighborhood.

Liquid Ester

The potential energy of cohesion for the melt of the solid ester was investigated at the melting point (91°C), at which the solid ester and the racemic liquid mixture of enantiomers coexist. The internal energy of the liquid is given by equation 27 in terms of internal energies for the solid ester and for melting. According to data in Figure 8 these terms amount to 708 and 33 kJ mol^{-1} which give a value of 741 kJ mol^{-1} (177 kcal mol^{-1}) for the internal energy of the liquid.

$$U_L = U_A + \Delta_M U \quad (27)$$

The kinetic energy for the liquid was then obtained. Since Debye function for entropy had been suitable for obtaining the vibrational energy of the gaseous ester, it was also used for the liquid. Liquid state resembles gas state in the free translation and rotation of molecules (each mode with $3RT/2$ energy [1]) although frequencies of molecular vibrations in the liquid differ from those in the gas as pointed out in Introduction (Fundamentals methods to determine potential energy of cohesion).

In a preliminary check to fit Debye vibrational entropy function to a tentative entropy of the liquid it was observed that the function accounted best for entropy when rotational motion was included [setting $N - 1$ rather than $N - 2$ in $S = (N - 1)f(T_D/T)$]. Accordingly, equation 28 was employed to work out the joint vibrational and rotational entropies at 91°C. This equation results from the experimental heat capacity of the liquid (equation 23) with subtraction of three degrees of freedom regarding translational motion ($dS = CdT/T$, $C_{V+R} = C - 3R/2$).

$$S_{V+R} = -0.056T^2 + 10.2T - 1570\ln T + 6270 + S_{V+R}(91°C) \quad (28)$$

Using equation 28 in the above variational manner regarding the gaseous ester, it gave an optimal S_{V+R}(91°C) of 346 J K^{-1} mol^{-1} with an extremely good fit of Debye function (Figure 9). The T_D value corresponding to this optimal entropy was 1410 K with an error as little as 0.01%. Finally, this T_D value was applied to the energy function to obtain the combined vibrational and rotational energies which added to the zero-point energy and the translational energy (3*RT*/2) gave a total kinetic energy of 743 kJ mol^{-1}.

The absolute entropy of the liquid ester (inclusive of the translational entropy) was obtained at 91°C from the known absolute entropy of the solid ester at 79°C by way of the change in entropy from 79 to 91°C and the entropy of melting at 91ºC ($\Delta S_M = Q_M/T$, Figure 8). The change in entropy for the temperature gap was obtained by the integrated heat capacity of solid ester (equation 9, $dS = CdT/T$). The absolute entropy of the liquid amounted to 452 J K^{-1} mol^{-1}, and includes the entropy of mixing of the enantiomers in the liquid state.

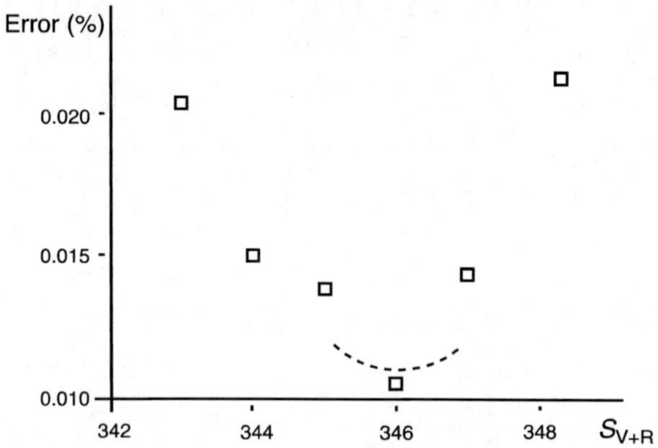

Figure 9. Maximal fitting to entropy for the liquid ester (S_{V+R} in J K^{-1} mol^{-1}).

From the worked-up total internal energy and kinetic energy of the liquid, a value of –2 kJ mol^{-1} (–0.5 kcal mol^{-1}) resulted for the potential energy of cohesion ($E_P = U - E_K$). As compared with the value for the potential energy of the solid ester or the polymorph (–11, –10 kcal mol^{-1}) the value may be low for a real value. It is possibly overlapped by an increase in internal electronic potential energy of binding of the molecule due to a higher dielectrical mobility in the liquid as pointed out in Introduction.

The involvement of rotational motion in Debye vibrational entropy function indicates that molecular rotation in the liquid is not free, in other words hindered rotations are actually vibrations. These hindered rotations in the liquid should arise from the nonspherical configuration of the molecule at the same time that cohesion forces correlate rotational motion between molecules in the liquid. It happens even though molecules undergo free translation and the liquid may be fully disordered.

Hindered molecular rotation in a liquid may be contrasted with free molecular rotation in a crystal such as a plastic crystal. Hindered molecular rotations in a liquid become lattice vibrations in a crystal, representing three degrees of freedom for lattice vibrations. The remaining free translational motion in a liquid becomes in a crystal the lattice vibrations due to oscillatory displacement of the molecular center of mass, which corresponds to the remaining three degrees of freedom of motion of the molecule.

The very high degree of agreement of Debye function for entropy with the entropy of the liquid is remarkable. It is higher than the agreement for the solid or gaseous ester, which is nevertheless adequate, and validates the function as a tool for determining in a simple way vibrational energies and entropies in the solid, liquid and gas states in conjunction with experimental or calculated data.

Table 3. Summary of data for the ester. Energies in kcal mol^{-1}

	Solid	**Gas**	**Polymorph**	**Liquid**
E_K	184, 179 (refined)	202, 196 (refined)	179	177
E_P	−11.2	0	−10.5	−0.5

Table 4. Absolute entropies for the ester (in cal K^{-1} mol^{-1})

	Solid	**Liquid**	**Gas**
S	81 (79°C)	108 (91°C)	164 (79°C)

There is not strong theoretical support for Debye functions in the liquid and gas states, not even for a molecular solid, as Debye's theory refers to monoatomic solids and to a continuous distribution of lattice frequencies. Appropriateness of Debye functions is otherwise of practical and analytical significance. Fitness of the energy function is almost as good as that of the entropy function and the former function may alternatively be used. The heat-capacity function, for which the theory was originally devised, is distinctively scanty for precision in the case of molecular solids.

CONNECTION OF POTENTIAL ENERGY TO TIME BY MEANS OF ENTROPY IN THERMAL TRANSFORMATIONS

As concerns a relationship of entropy to time the background consists of the principle of equivalence of energy $dU = TdS - pdV$, the definition of entropy $dS = dQ/T$, and a set of circumstantial functions in the variables U, T, S, p and V, basically heat TS and work pV. These circumstantial functions are pertinent depending on the particular process. The variables in these merely assisting functions are mathematically correlated by Maxwell equations forming a thermodynamic frame of operation consistent with the preceding principles, and therefrom connected to other physically consistent thermodynamic parameters. Mechanical work in chemical transformations, pdV, can be generalized to other forms of work, Ydx.

Time is classically introduced in thermodynamics by a division of the two sides of a thermodynamic equality by time. This is a dependable introduction of magnitude time, as this is compelled into an equality. This artificial introduction of time does not eradicate the modesty of thermodynamics to time as the contingency of time is maintained. Moreover, it sometimes results in time coefficients having an awkward meaning or leads to intricate statistical kinetics. It will be shown in the following that time can be introduced into just one side of a thermodynamic equation, what represents an undependable and genuine introduction of magnitude time into the body of thermodynamics. Moreover, the particular underlying equations are general.

The condition required for bringing time is that the system is thermodynamically isolated, in other words the system is thermally isolated (an adiabatic system, $Q = 0$) and does not produce or receive work (an isochoric system, $W = 0$). It will be become obvious in this section that time cannot be introduced in an unambiguous manner for a working system. The resultant function that correlates entropy and time is valid for any thermodynamically isolated system regardless of its nature or complexity.

For such a thermodynamically isolated system, entropy increases in a process according to second law. In the system, internal energy remains constant during the process ($\Delta U = Q - W$, $Q = W = 0$). In order that the principle of equivalence of energy can be handled it is required that the increase in entropy in the system is taken as an *internal* heat, what is entirely reasonable due to the actual increase in entropy taking place in the process ($dQ = TdS > 0$).

As regards work, *internal volume* is to be analogously used as a variable parameter for handling the principle of equivalence since the volume of the system is constant ($dW = pdV_0$). Such needed instrumental internal volume is however not an artificial physical concept, but is physically correct in substance because the *internal work* that is connected to it is here analogous to internal heat which is a physically substantiated concept according to second law. Also, the actual internal heat must convey an equivalent internal work so that the internal energy of the system remains constant as it is. Moreover, internal volume is a concept alike to the internal pressure of cohesion pointed out in Introduction, which is a sound concept.

The key equation correlating entropy and volume for the thermodynamically isolated system (equation 29, V_I internal volume) results by elemental differential-calculus operations from the principle of equivalence applied to the system, $TdS - pdV_0 = 0$, from the pertinent assisting function, $U = TS - pV_0 = 0$, and Maxwell equation for the system, $dp/dT = dS/dV_I$. It can be demonstrated that latter equation holds for the system [26].

$$\frac{dS}{S} = \frac{dV_I}{V_0} \tag{29}$$

Correctness of equation 29 is not obvious from the viewpoint of differential calculus as it simultaneously refers to a constant volume and its differential variable. Moreover, the more general equation 30 is correct for a thermally isolated system not having the restriction of volume constancy. It may thus seem that equation 29 is an artifact from 30. However, it can be rigorously and exhaustively demonstrated that equation 29 is mathematically correct [26]. The more general equation 30 holds for a thermally isolated system (an adiabatic system) undergoing a limited or free expansion.

$$\frac{dS}{S} = \frac{dV}{V} \tag{30}$$

Having achieved equation 29, the basic differential equation which correlates entropy and time, equation 31, ensues in a straightforward manner by means of an adimensional substitution of magnitude time for magnitude volume at the adimensional equation 29. Quantity k in equation 31 is a constant that replaces constant $1/V_0$. The former has the units of a first order rate constant (τ^{-1}) and is characteristic to each process.

$$\frac{dS}{S} = kd\tau \tag{31}$$

Time cannot be inserted in an unambiguous manner but for the exclusive entropy-volume equation 29. Thus, when replacement by time is made on equation 30 (which refers to other conditions) the constancy of a velocity constant cannot be established because denominator V is not a constant, to say $k = 1/\tau \neq$ const. versus $k = 1/\tau_0$. Then, the relationship of entropy to time can be any function [$dS/S = xd\tau$, $S = f(\tau)$] and the insertion of time is actually not genuine for ambiguous.

The novel present way to introduce time is not quite the same that the common way of making up time from a division at both sides of an equation as in the way being now considered entropic side *dS/S* remains intact and undependable from time. In this way, time by itself becomes an independent and genuine magnitude of thermodynamics.

Integration of equation 31 between limits provides an explicit function correlating entropy to time (equation 32). In this function the final entropy is asymptotic to time. It suggests that the time required for reaching thermodynamic equilibrium in a transformation or for completing a process is infinite. However, this is not congruent with empirical knowledge that indicates a finite time for a natural transformation to take place. Besides, it springs from second law the notion that, in a thermally isolated system, a maximal entropy is attained when thermodynamic equilibrium is reached in a temporal process [7, 22]. Function 32 lacks, on the contrary, a maximum point. Therefore the function, being a genuine entropy-time function as it is, is to be somehow modified so that conforms such demands.

$$S = S_2 - (S_2 - S_1)\mathrm{e}^{-k\tau} \tag{32}$$

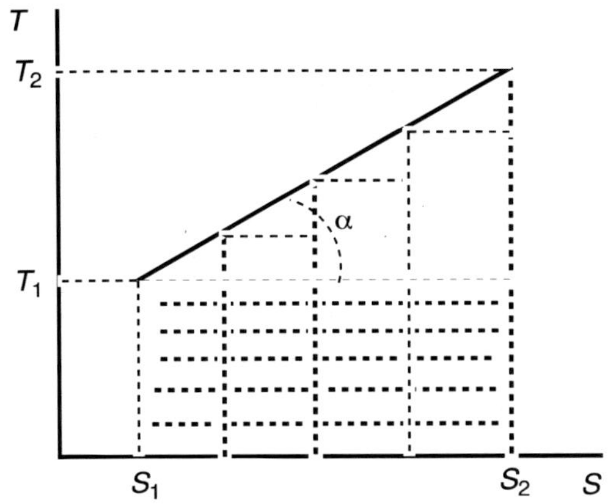

Figure 10. Irreversible heat ($dQ = TdS$).

The key to a maximal entropy in the entropy-time relationship is the principle that, in an irreversible transformation, heat is less that in the corresponding reversible transformation, as enunciated in second law. This is illustrated in the temperature-entropy diagram of Figure 10, in which heat in a reversible process is the total area under the line while the corresponding heat in the irreversible process is the shaded area. In a partially irreversible process, heat is the drawn up area, which depends on the degree of irreversibility for an actual process. Temperature increases in a process in a thermodynamically isolated system, and the reference temperature for evaluating the irreversible internal heat produced is the initial temperature ($Q = T_1\Delta S$), like an isothermal transformation.

Internal heat is generated against a higher temperature, but it does not contradict the principle that heat is transferred from higher to lower temperature as internal heat and not external heat is here concerned. Nevertheless, the reference temperature for valuating external heat in heat transfers that take place in an irreversible manner is the lower end temperature since the principle of minimal irreversible heat would be otherwise contradicted. It is similar to work in an irreversible expansion in which the amount of work produced is determined by the lower final pressure.

Function 32, resulting from a differential equation as it does, refers to a reversible path parallel to the actual irreversible path, and thus misses the thermodynamic requisite of minimal irreversible heat as illustrated in Figure 10. The reversible path in entropy-time function 32 corresponds in entropy-temperature function to the slanted line in Figure 10. This line is to be rotated by angle α in order that the actual irreversible path is in conformity with the irreversible heat.

So this rotation was applied to equation 32 for satisfaction of the requirement. The transformation equations for the rotation are 33 and 34 [$\alpha = \arctan(T_2 - T_1)/(S_2 - S_1)$]. A complete transformation with regard to the S-τ coordinate axes involves three additional geometrical operations for a perfect match of the transformed S-τ function to actual irreversible heat. The resultant is implicit function 35, in which parameter a, b and c are expressed in terms of rate constant k, and T_1, T_2, S_1 and S_2. Equation 35 is the mathematical function to the entropy-time relationship for an irreversible process in a thermodynamically isolated system.

$$\tau = \tau'\cos\alpha - S'\sin\alpha \qquad (33)$$

$$\tau = \tau' \sin\alpha + S' \cos\alpha \tag{34}$$

$$S + (\tau + a)\sin\alpha\cos\alpha = b\cos\alpha\{1 - e^{k[S\tan\alpha - (\tau + a)\cos\alpha]}\} + c \tag{35}$$

The exponential curve, $y = e^x$, provides maximum points by rotations of the coordinate axes as the asymptote at infinity converts into maxima of the curve. The maximal entropy at finite time that comes about in curve 35 by the rotation of curve 32 provides the maximal entropy at thermodynamic equilibrium, as advanced by second law for a thermodynamically isolated system in irreversible evolution.

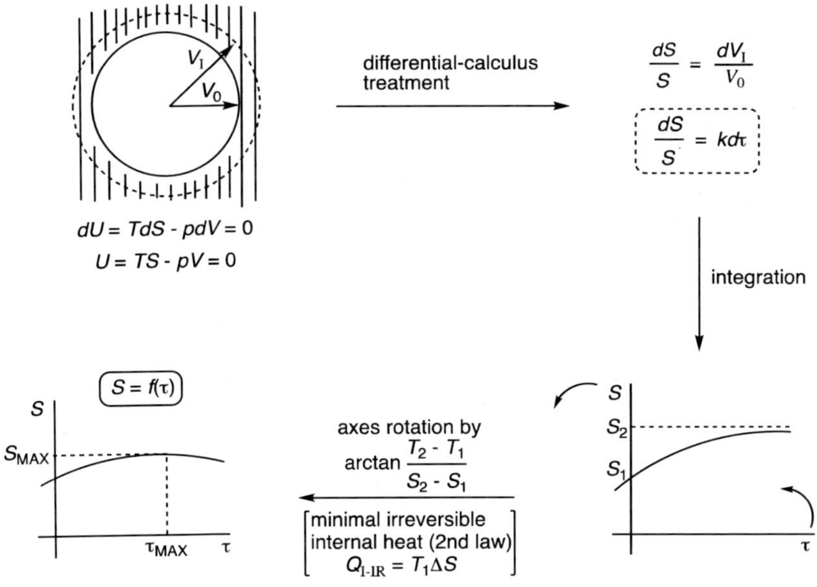

Figure 11. Path to entropy-time relationship (V_I internal volume).

The stepwise procedure described is depicted in Figure 11. Both entropy and time are contracted upon the transformation, entropy from the final value in the reversible process to the maximal entropy in the irreversible process, and time from infinity to the finite time corresponding to maximal entropy and thermodynamic equilibrium. Also, there occurs a collateral reduction of the gap between initial and final entropies as well as between initial and final temperatures, but the main point is the uncovering of a maximal entropy at a finite time.

INVOLVEMENT OF POTENTIAL ENERGY IN ENTROPY-TIME RELATIONSHIP

Concerning work of potential energy in the thermodynamically isolated system considered, it can be expressed by the pressure of cohesion due to internal cohesion forces and by the internal volume, in the manner pointed out in Introduction with respect to noninternal volume (equation 5). Pressure of cohesion is a negative quantity and internal volume increases in the process here according to equation 29 ($dS > 0$), thus the work of potential energy is negative.

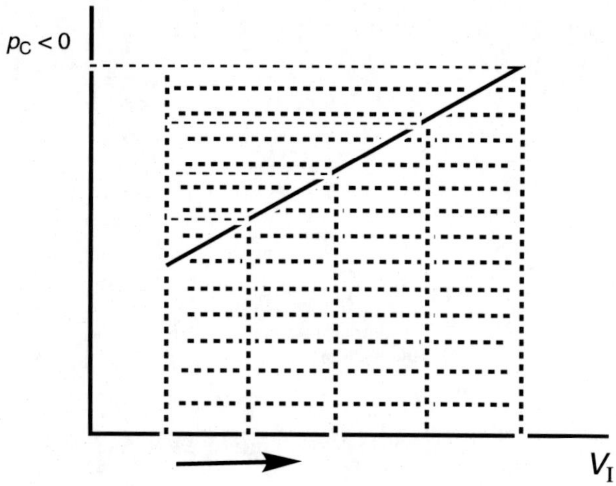

Figure 12. Irreversible work of potential energy ($W = p_C\Delta V_I < 0$).

In Figure 12, irreversible work of potential energy is represented by the total shaded area, which refers to the final cohesion pressure, similarly to the irreversible mechanical work in an expansion although the final pressure in the present case is higher than the initial pressure. This inversion in the reference pressure is due to the negative character of cohesion presssure contrarily to external pressure. For a contraction in internal volume, lower final cohesion pressure is the reference for irreversible work, and this will be positive.

As regards internal energy in the process ($\Delta U = 0$), a negative potential-energy work amounts to a negative internal heat ($W_I = Q_I$), and the latter thus contributes in a negative way to the overall increase in entropy in the irreversible process. This entropic effect of potential energy is connected with the spacial entropy mentioned in Introduction (Entropy in liquids). Potential energy represents a negative entropic contribution to the spacial entropy that arises from the increase in internal volume in the irreversible process. If the process occurred in a reversible manner, entropy would be constant according to second law. Under those conditions field forces would act in an equilibrium manner and potential energy would not give rise to either internal heat or entropy.

As it has been above shown, the actual entropy-time correlation rises by agency of angle α [$\alpha = \arctan(T_2 - T_1)/(S_2 - S_1)$, Figure 10]. This angle means that attainment of a maximal entropy with a reduction in the time required for ending the irreversible process is facilitated by the change in temperature in the process. Thus the larger the temperature increase the larger the angle α. Then the reduction of time for ending the process in attaining equilibrium at maximal entropy is exalted by a larger α. As concerns denominator $S_2 - S_1$, a lower final entropy favors attainment of the equilibrium, in contrast with temperature in the numerator.

The real activity of potential energy in the process refers to the direction whereby potential energy is decreased, losing its power to do any action and losing its identity. In other words, a contrary increase in potential energy may be promoted by collateral thermal factors in the process and does not pertain to the inner role of potential energy. Loss of potential energy in the irreversible process conveys a negative entropy change as it has been above pointed out. This negative entropy lowers the final entropy at the denominator of $\tan\alpha$, and α accordingly increases. Consequently, the effect of potential energy is to favor attainment of thermodynamic equilibrium with a reduction of the time required. This effect of potential energy in the thermodynamically isolated process by means of entropy is neat as the entropy is unmixed with temperature in tanα. This is so because spacial entropy is independent of temperature as it has been pointed out in the chapter.

CONCLUSION

Potential energy of cohesion can be determined with precision from caloric measurements together with elemental statistical mechanics with simple variational procedures. Potential energy brings about a reduction of time for fullness of irreversible unassisted thermodynamic processes.

REFERENCES

[1] Díaz Peña, M., and Roig Muntaner, A. 1984. *Química Física* [*Physical Chemistry*]. Madrid: Alhambra.

[2] Ingold, C. K. 1969. *Structure and Mechanism in Organic Chemistry.* Ithaca: Cornell University Press.

[3] Hine, Jack. 1975. *Structural Effects on Equilibria in Organic Chemistry.* New York: John Wiley.

[4] Morse, P. M. (1929). Diatomic Molecules According to the Wave Mechanics. II. Vibrational Levels. *Physical Review*, 34: 57-64.

[5] Herzberg, Gerhard. 1950. *Spectra of Diatomic Molecules.* New York: Van Nostrand Reinhold.

[6] Debye, P. (1912). Zur Theorie der spezifischen Wärmen [On the Theory of Specific Heats].. *Annalen der Physik*, 39: 789-839.

[7] Aguilar Peris, José. 1970. *Termodinámica y mecánica estadística.* [*Thermodynamics and Statistical Mechanics*]. Valencia: Saber.

[8] Einstein, A. (1907). Die Plancksche Theorie der Strahlung und die Theorie der spezifischen Wärme [The Planck Theory of Radiation and the Theory of Specific Heat]. *Annalen der Physik*, 22: 180-190.

[9] Nernst, W., Lindemann, F. A. (1912). Spezifische Wärme und Quantentheorie [Specific Heat and Theory of Quanta]. *Zeitschrift für Elektrochemie*, 18: 817-827.

[10] Herzberg, Gerhard. 1945. *Infrared and Raman Spectra of Polyatomic Molecules.* New York: Van Nostrand Reinhold.

[11] Eliel, Ernest L. 1962. *Stereochemistry of Carbon Compounds.* New Yok: McGraw-Hill.

[12] Jacques, Jean, Collet, André, and Wilen, Samuel H. 1994. *Enantiomers, Racemates, and Resolutions.* Malabar: Krieger.

[13] Mislow, Kurt. 1965. *Introduction to Stereochemistry.* Reading: Benjamin/Cummings.

[14] Ros, F., Barbero, I. (2005). Formation of Tertiary 3-Nitro Allylic Alcohol by Condensation of 2-Chloroisobutyrophenone with Nitromethanide Anion: Estimation of Gibbs Free Enthalpies of Reaction in Solution for the Reaction Mechanism. *Monatshefte für Chemie – Chemical Monthly*, 136: 1607-1617.

[15] Sabbah, R., Xu-Bu, A., Chickos, S., Planas Leitao, M. L., Roux, M. V., Torres, L. A. (1999). Reference materials for calorimetry and differential thermal analysis. *Thermochimica Acta*, 331: 93-204.

[16] Colomina, M., Jiménez, P., Turrión, C. (1982). Vapour pressures and enthalpies of sublimation of naphtalene and benzoic acid. *Journal of Chemical Thermodynamics*, 14: 779-784.

[17] Callanan, J. E., Sullivan, S. A. (1986). Development of standard operating procedures for differential scanning calorimeters. *Review of Scientific Instruments*, 57: 2584-2592.

[18] Landolt-Börstein, *Zahlenwerte und Funktionen* [*Numerical Values and Functions*]. 1961. Vol. 2. Berlin: Springer.

[19] Davidson, Norman. 1962. *Statistical Mechanics.* New York: McGraw-Hill.

[20] Rihani, D. N., Doraiswamy, L. K. (1965). Estimation of Heat Capacity of Organic Compounds from Group Contributions. *Industrial & Engineering Chemistry Fundamentals*, 4:17-21.

[21] Perry, R. H., and Chilton, C. H., eds. 1973. *Chemical Engineering's Handbook.* New York: McGraw-Hill.

[22] Denbigh, Kenneth. 1971. *The Principles of Chemical Equilibrium.* London: Cambridge University Press.

[23] Hirschfelder, J. O. (1940). Simple Method for Calculating Moments of Inertia. *Journal of Chemical Physics*, 8: 431.

[24] Wilson, E. B. (1940). The Present Status of the Statistical Method of Calculating Thermodynamic Functions. *Chemical Reviews*, 27: 17-38.

[25] Roux, M. V., Jiménez, P., Vacas, A., Cano, F. H., Apreda-Rojas, M. C., and Ros, F. (2003). A Compact and Lipophilic Enantiomer Showing a Bilayer Crystal Structure – Free Energy of Spontaneous Resolution of the Racemic Compound into the Crystalline Enantiomers. *European Journal of Organic Chemistry*, 2003: 2084-2091.

[26] Ros, Francisco. *Undisclosed results.*

In: Advances in Energy Research. Volume 31 ISBN: 978-1-53616-271-4
Editor: Morena J. Acosta © 2019 Nova Science Publishers, Inc.

Chapter 6

A REVIEW ON POWER GENERATION FROM SALINITY GRADIENT ENERGY BY OSMOSIS

Yanmei Jiao[1] and Chun Yang[2,*]
[1]School of Physical and Mathematical Sciences,
Nanjing Tech University, Nanjing, Jiangsu, China
[2]School of Mechanical and Aerospace Engineering,
Nanyang Technological University, Singapore

ABSTRACT

Salinity gradient energy is referred to as the chemical potential difference between two solutions with different concentrations. It commonly occurs at river mouths where a river flows into a sea/ocean and it has an equivalent hydraulic head as high as 270 m, which is also termed as a 'silent waterfall'. Deterioration in environment and depletion of conventional energy resources arise the global need in finding alternative energy sources. The salinity gradient energy, as the second largest marine-based energy source and one of the most potential energy resources, has attracted an increasing attention over the past decade. Osmosis is considered as a potential way to harvest this kind of salinity gradient

* Corresponding Author's E-mail: mcyang@ntu.edu.sg.

energy. Extensive research works have been conducted to develop green and environment-friendly power generation methods based on osmosis-related mechanisms, and they include pressure-retarded osmosis (PRO) method, forward osmosis-electrokinetic (FO-EK) method, and diffusio-osmotic (DO) method. This chapter provides an overview of these three methods for sustainable power generation from the salinity gradient energy by using osmosis based processes.

Keywords: salinity gradient energy, pressure retarded osmosis, forward osmosis, diffusio-osmosis, green power generation

1. Introduction

Nowadays, the need for alternative environmental friendly 'green energy' has long been a hot issue and becomes even more important due to deterioration in environment and depletion of traditional fossil energy resources. Salinity gradient energy, commonly occurring at river mouths where a river flows into a sea/ocean, is one of the green energies due to its huge amount and non-interference from environmental variation (Ramon, Feinberg, & Hoek, 2011; Rankin & Huang, 2016; Yeh, Chen, Chiou, & Su, 2017). Salinity gradient energy is referred to as the chemical potential difference between two solutions with different concentrations, and is equivalent to osmosis gradient energy. Various kinds of methods have been developed in bench and large scales to harvest this kind of salinity gradient energy. They include the conventional methods, such as the pressure retarded osmosis (PRO) (Sidney Loeb, 1976), the reverse electrodialysis (RED) (Pattle, 1954), and the vapor pressure difference (VPD) (Olsson, Wick, & Isaacs, 1979), and the emerging methods such as the electric double-layer capacitor (EDLC) (Brogioli, 2009), the faradaic pseudo-capacitor (FPC) (La Mantia, Pasta, Deshazer, Logan, & Cui, 2011), the forward osmosis-electrokinetic (FO-EK) (Jiao, Yang, & Kang, 2014), and the diffusio-osmosis (DO) (Siria et al., 2013). Among all these methods, the PRO, FO-EK, and DO three methods have a common feature that their driving forces applied in generating power fully or partially come from the mechanism of osmosis. Osmosis in general is defined as the net transport of solvent between two solutions with different concentrations separated by a semipermeable membrane which allows the permeation of the small solvent molecules but blocks the large solute molecules or ions (Jiao, Kang, & Yang, 2015). The difference in solution concentration leads to a chemical potential difference (i.e., the osmotic pressure difference) between the two solutions. Such chemical potential difference in essence is the driving force for transport of solvent molecules from the solution of high concentration side to that of the low concentration side. Specifically, water is exclusively used as the solvent in these three osmosis-based power generation processes.

This chapter reviews these three osmosis-based methods in harvesting the salinity gradient energy, not only providing their state-of-the-art development but also highlighting their pros and cons. The challenges and optimizing strategies as well as future development are outlined.

2. Fundamentals of Osmosis

The solvent permeating direction of the osmosis process across a membrane is dependent on two factors: the inherent osmotic pressure difference, $\Delta\pi$, and the externally applied pressure difference, ΔP. As a rule of thumb, the solution where the solvent permeates from/into the membrane system is termed as the feed solution (FS)/the draw solution (DS), respectively. The osmotic pressure of a solution with its concentration c is given by the osmotic equation (i.e., Eq. (1)), and such equation was derived based on the theory of chemical equilibrium with two main assumptions: (1) the solution of single component is separated from the corresponding pure solvent by a semipermeable membrane, and (2) the osmosis system is in its equilibrium state. Parameters of χ, R, T, and v in the osmotic equation are the osmotic coefficient, the universal gas constant, the temperature and the stoichiometric number respectively, and the subscript 'a' and 'c' respectively represent the anion and the cation (Strathmann, 2004). Clearly, this equation shows that the osmotic pressure linearly decreases with increasing the solution concentration. Then the inherent osmotic pressure difference is determined by the difference between the inherent osmotic pressures of the DS and the FS, i.e., $\Delta\pi = \pi_{DS} - \pi_{FS}$. Meanwhile, the externally applied pressure difference is the difference between two pressures acting on the FS and the DS, i.e., $\Delta P = P_{FS} - P_{DS}$. The subscript 'FS' and 'DS' represent the feed solution and the draw solution, respectively.

$$\pi = -\chi \mathrm{RT}(\nu_a + \nu_c)c \tag{1}$$

The solvent permeating flux, J_w, is given by (K. L. Lee, Baker, & Lonsdale, 1981):

$$J_w = A(\Delta\pi - \Delta P) \tag{2}$$

where A is the solvent permeability coefficient. Consequently, three different osmosis situations, including the forward osmosis (FO), the pressure retarded osmosis (PRO) and the reverse osmosis (RO), can be classified according to the different relationships between $\Delta\pi$ and ΔP. For FO, no pressure difference is applied and the solvent permeates from the low concentration side to the high concentration side. For PRO, although the applied pressure acting on the DS side is higher than that acting on the FS side, the absolute value of ΔP is smaller than that of $\Delta\pi$. Therefore, the direction of solvent permeation in the PRO process is the same as that in the FO process. For RO, since the applied pressure acting on the DS side is much smaller than that acting on the FS side and the absolute value of ΔP is larger than that of $\Delta\pi$, the solvent permeating flux is from the high concentration side to the low concentration side. Usually, FO and PRO are used for power generation, while RO is employed in water desalination.

The performance of osmosis processes is commonly degraded by three main factors which are fouling, scaling, and concentration polarization (CP) (S. Lee, Boo, Elimelech, & Hong, 2010; Zhang et al., 2012). Generally, the first two factors happen on or within the membrane and the last one occurs in the solution nearby the membrane region. The membrane fouling is caused by accumulation of suspending organic matters onto the surface or into the inner structure of the membrane. The membrane scaling is resulted from precipitation of inorganic minerals, such as calcium, magnesium, barium, bicarbonate, and sulfate as well as deposition of colloidal particles, onto the membrane. Both fouling and scaling ultimately lead to degradation of the solvent permeating flux during osmotic processes. However, the thickness of the fouling and scaling layers during FO and PRO processes is considerably different from that of the RO process because of different operation modes. Meanwhile, chemical methods are often used to remove the membrane fouling and scaling to regain the solvent permeating flux in the RO process, while during FO and PRO processes, the recovery is achieved by increasing the velocity of cross-flow. As for the CP, although it happens within the solution, it is strongly related to the membrane asymmetric structure, which practically consists of a thin and dense selective layer supported by a thick and porous polymer layer. Both the solute and the solvent molecules can freely pass through the porous supporting layer, but only the solvent molecules can pass through the dense selective layer. Therefore, two kinds of CP, including the external concentration polarization (ECP) happened adjacent to the membrane and the internal concentration polarization (ICP) happened inside of the porous supporting layer of the membrane, are considered in osmotic processes. Since high external pressures are usually applied in RO, the effect of ICP on the RO performance, compared to the overwhelming effect of ECP, is small and thus can be reasonably ignored. Therefore, only the ECP is considered in the mathematical modeling of the solvent permeating flux in RO, and the ECP can be derived from the following equation:

$$J_w = A\left[\pi_{\mathrm{DS,b}}\exp\left(-\frac{J_w}{k_{\mathrm{DS}}}\right) - \pi_{\mathrm{FS,b}}\exp\left(-\frac{J_w}{k_{\mathrm{FS}}}\right) - \Delta P\right] \tag{3}$$

where the subscript ‘b’ represents the bulk region of the corresponding FS/DS, and k is the mass transfer coefficient given by $k=Sh\ D/D_h$ with Sh, D, and D_h being the Sherwood number, the solute diffusivity, and the hydraulic diameter, respectively. The term, $\pi_{\mathrm{DS,b}}\exp(-J_w/k_{\mathrm{DS}})-\pi_{\mathrm{FS,b}}\exp(-J_w/k_{\mathrm{FS}})$, can be regarded as the effective osmotic pressure difference $\Delta\pi_{eff}$

across the membrane during the RO process. As for analyzing the solvent permeating flux in the FO and PRO processes, the model development should take into account both the ICP and the ECP effects. Moreover, different orientations of the asymmetric membrane give rise to different types of ICP/ECP. When the dense selective layer of the membrane faces against the DS, commonly being regarded as the normal mode, the solutes enrich within the porous supporting layer but dilute near the dense selective layer at the DS side. This leads to the concentrative ICP and the dilutive ECP. On the contrary, when the porous support layer faces against the DS, commonly being regarded as the reverse mode, the solutes enrich near the dense selective layer at the DS side and dilute within the porous supporting layer. This leads to the concentrative ECP and the dilutive ICP. Then the general forms of the solvent permeating flux for FO/PRO in the normal mode and the reverse mode are expressed by Eq. (4) and Eq. (5), respectively. The parameter S in those two equations denotes a structure factor related to the porous supporting layer of the membrane, which is defined as $S=t\tau/\vartheta$ with t, τ, and ϑ representing the thickness, the tortuosity, and the porosity of the porous supporting layer, respectively. The terms, $\pi_{DS,b}\exp(-J_w/k_{DS})-\pi_{FS,b}\exp(-J_wS/D)$ and $\pi_{DS,b}\exp(-J_wS/D)-\pi_{FS,b}\exp(-J_w/k_{FS})$, are the effective osmotic pressure differences of FO/PRO across the membrane. Obviously, for FO, the externally applied pressure difference is zero, i.e., $\Delta P = 0$, and, for PRO, the absolute value of the effective osmotic pressure difference is larger than that of the applied pressure difference, i.e., $|\Delta\pi_{eff}| > |\Delta P|$ (Sidney Loeb, 1976). From a practical viewpoint, the presence of CP reduces the performance of the osmotic processes due to the degraded effective osmotic pressure difference between the FS and the DS. In FO/PRO applications, the normal mode is frequently employed since the CP phenomenon in the normal mode is less significant than that in the reverse mode, and thus it leads to a relatively higher effective osmotic pressure difference.

$$J_w = A\left[\pi_{DS,b}\exp\left(-\frac{J_w}{k_{DS}}\right) - \pi_{FS,b}\exp\left(-\frac{J_w S}{D}\right) - \Delta P\right] \tag{4}$$

$$J_w = A\left[\pi_{\mathrm{DS,b}} \exp\left(-\frac{J_w S}{D}\right) - \pi_{\mathrm{FS,b}} \exp\left(-\frac{J_w}{k_{\mathrm{FS}}}\right) - \Delta P\right] \tag{5}$$

Along with the osmotic processes, the solutes can unavoidably diffuse across the semipermeable membrane in practical applications due to membrane's non-ideal characteristic. The permeating property of solutes through a semipermeable membrane is reflected by the so-called solute rejection, $R_j = 1 - c_p/c_{\mathrm{FS}}$, where c_p is the concentration of the permeate solute, and the term c_p/c_{FS} denotes the solute passage.

Generally, the membrane and the FD/DS flow patterns are considered as two critical aspects for optimal performance of osmotic processes. The semipermeable membranes with properties such as high salt rejection, high water permeability, and thin porous supporting layer, are preferred. Meanwhile, the FS/DS flow patterns greatly influence the effects of fouling, scaling and CP.

3. Osmosis-Related Power Generation Methods

3.1. Forward Osmosis-Electrokinetic (FO-EK) Method

As aforementioned, FO can provide considerable driving forces for the process of power generation. Figure 1 shows one of our previous works proposed a novel FO-EK power generation system to harvest the salinity energy (Hon, Zhao, Yang, & Seow, 2012). As implied by the name, the hybrid system mainly encompasses two phenomena, i.e., FO and EK. The EK phenomenon commonly happens at the interfaces of majority of objects, such as the porous glass shown in Figure 1 (a), which can be reasonably charged at the interface once being immersed in a liquid through electrochemical mechanisms such as adsorption of charged species and dissociation of ionizable groups (Everett, 1988; Hunter, 1981). The structure of the porous glass can be equivalently regarded as a bundle of microchannels packed in parallel. One typical microchannel is illustrated in

Figure 1 (b). The channel surface is negatively charged due to the dissociation of the silanol group when in contact with an electrolyte solution, and thus it attracts cations and repels anions in the electrolyte solution, leading to the formation of an electric double layer (EDL). The electrolyte solution within the microchannel exhibits an overall positively-charged electric property. When a pressure driven flow passes through the microchannel, a streaming potential is induced by accumulation of the cations at the down streaming terminal and, simultaneously, a streaming current is generated by the forward convection flow. The streaming potential and the streaming current can be properly harvested as a kind of power source when the two ends of the microchannel are connected by an external electric circuit. Rather than employing an external pump to generate the pressure, the proposed FO-EK system, shown in Figure 1 (c), utilizes a FO sub-module as the driven force unit to induce the convection flow of the FS passing across the porous glass in the EK sub-module. The electric energy is thus produced by the EK sub-module. The overall system can realize the power generation process from the salinity energy with no mechanical energy input.

Theoretically, Hon et al., treated the porous glass as a bundle of microchannels with uniform radius, and developed a semi-empirical mathematical model for the whole FO-EK power generation system. However, it has been shown that the radius indeed plays an essential role in the energy conversion performance of the pressure driven EK flow. The energy conversion efficiency reaches the maximum when the critical microchannel radius is around two times of the EDL thickness, and it decreases sharply when the microchannel radius deviates from the critical microchannel radius (Chein, Liao, & Chen, 2009). Realizing the non-uniform structure of the porous glass, Jiao et al., refined the mathematical model in one of our follow-up works using an assumption that the radii of the equivalent microchannels of the porous glass follow the log-normal probability density distribution. Accordingly, the refined model yields much better agreement with the experimental results (Jiao, Yang, & Kang, 2014). Furthermore, the power density is defined as the ratio of the generated power to the effective overall volume of the porous glass. It is assumed the streaming current linearly decreases with the increment of the streaming potential, then the maximum power happens at the half of the streaming potential, i.e., (Chang & Yang, 2009; Daiguji, Yang, Szeri, & Majumdar, 2004):

$$W_{\text{FO-EK,max}} = \frac{1}{4}\frac{\Delta\phi I_s}{V_{eff}} \quad (6)$$

where $\Delta\phi$ and I_s are respectively the streaming potential and the streaming current, and the V_{eff} is the effective volume of the porous glass used in the EK sub-module. Experimentally, the maximum power density of 0.165 W/m^3 was achieved when the average pore size of the porous glass was 6 μm and the salinity difference between the FS and the DS was 4 M. Compared to the existing model that gives an over-estimation of the power generation density by 212%, the refined model predicts only 20% higher than the experimental results (Jiao, Yang, & Kang, 2014).

In addition to the refinement in the theoretical model of the FO-EK system, experimental improvements were also achieved. It is known that a more hydrophobic interface and a larger surface charge density of the microchannels give rise to a faster liquid flow and a higher zeta potential, respectively, resulting in greatly enhanced energy conversion efficiency of the EK flows (Bouzigues, Tabeling, & Bocquet, 2008; Einzel, Panzer, & Liu, 1990; L. Joly, Ybert, Trizac, & Bocquet, 2006; Laurent Joly et al., 2004). However, those two factors usually do not occur simultaneously, meaning that the enhancement of the one factor weakens the other (Eijkel, 2007). In our follow-up work, both ultrasonic physical treatment and Sodium dodecyl sulfate (SDS) surfactant treatment (with a SDS concentration of 12 mM) of the porous glass were used to enhance the electrokinetic energy conversion performance. It showed that the maximum power density of 3.08 W/m^3 was obtained, and it was around 30% higher than that generated by the same untreated porous glass (Jiao, Yang, Kang, & Hon, 2014).

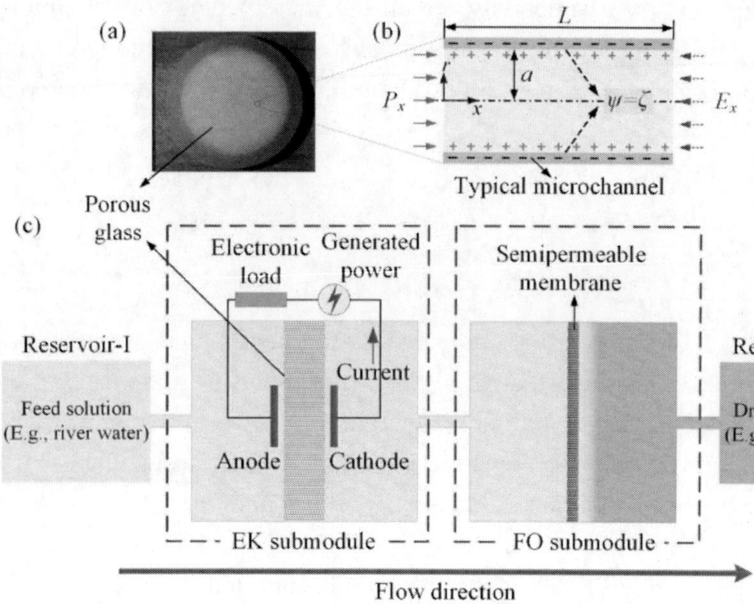

Figure 1. (a) Porous glass used in the forward osmosis-electrokinetic (FO-EK) power generation system to provide a bundle of microchannels for EK flows. (b) A pressure driven EK flow passing through a typical equivalent microchannel of the porous glass. The microchannel surface is negatively charged due to the dissociation of silanol groups, which attracts cations and repels anions in the electrolyte solution. The symbols P_x and E_x respectively represent the applied pressure gradient $-dP/dx$ and the induced electric field strength $-\partial\phi/\partial x$, and a, L, and ζ are the radius, the length, and the zeta potential of the microchannel, respectively. (c) Schematic of the FO-EK power generation system. Adapted with permission from (Jiao, Yang, & Kang, 2014). Copyright (2014) American Chemical Society.

3.2. Pressure Retarded Osmosis (PRO) Method

PRO power generation from salinity gradient energy was first reported by Loeb (1976), and Figure 2 shows the schematic of a continuous PRO method based power generation system. The system however suffers several drawbacks: (1) low efficiency of rotating components, (2) friction losses in plant streams, and (3) membranes not perfectly semipermeable. Hollow-fibers enclosed in minipermeators (Permasep B-10) in the form of a 'U' were utilized as the semipermeable membrane. The feed brine was pressurized and filtered outside the space of the hollow fibers while the water permeated into the inside space. The water permeation flux was determined by the flow rate difference between the permeat-donor and the flushing solution. The continuous permeation of water was from the water side to the brine side because the presence of the osmotic pressure increased the pressure of the brine solution, thereby driving the hydroturbine and the generator to generate electricity. Part of the generated electricity was used to power the water pump and the recycle pump, and the remnant was merged into the busbar (Sidney Loeb, 1976) (Sidney Loeb, Van Hessen, & Shahaf, 1976).

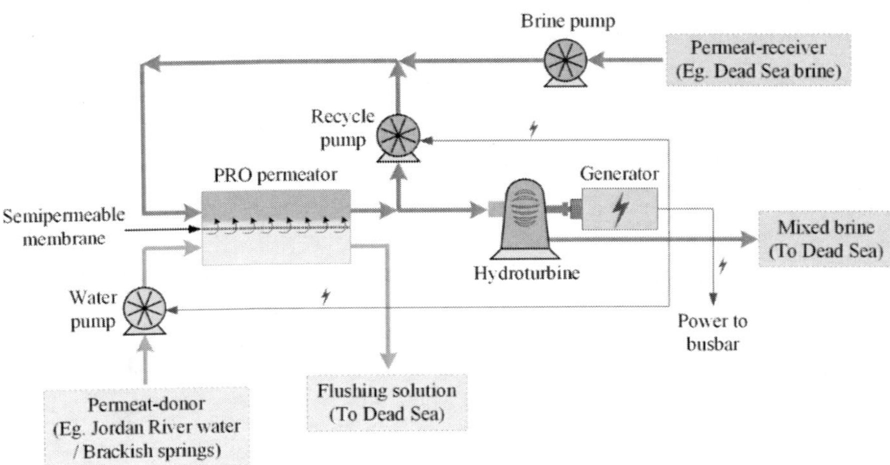

Figure 2. Schematic of a pressure retarded osmosis (PRO) method based power generation plant located in Dead Sea. Figure adapted from Ref. (Sidney Loeb, et al., 1976).

In the PRO power generation process, the power density is defined as the generated power per unit membrane area, and it is expressed by the product of the water permeating flux and the hydraulic pressure difference across the membrane:

$$W_{\text{PRO}} = A(\Delta\pi - \Delta P)\Delta P \tag{7}$$

Then the maximum power density occurs when $\Delta P = \Delta\pi/2$, and hence (K. L. Lee, et al., 1981):

$$W_{\text{PRO,max}} = A\frac{\Delta\pi^2}{4} \tag{8}$$

3.3. Diffusio-Osmosis (DO) Method

As for the membrane-based power generation methods, low separation efficiency of the membrane is the main barrier in achieving high energy conversion performance from salinity gradient energy. In recent years, great efforts have been made on exploring novel mechanisms in converting the salinity gradient energy into electricity. An emerging method termed as diffusio-osmosis (DO) has been developed for harvesting the salinity gradient energy. DO is referred to as an electrokinetic phenomenon happening within a few nanometers region of a charged nanochannel wall, and shown in Figure 3 (a) is such region characterized by the Debye layer λ_D. In the literature, the DO phenomenon describing of the motion of a surface charged colloid driven by an imposed concentration gradient has been studied, and the corresponding slip velocity was derived (Anderson, 1989; Prieve, Anderson, Ebel, & Lowell, 1984). During the DO method based power generation process, the application of concentration gradient along the nanochannel axial direction induces an interfacial diffusion-osmotic flow. The DO velocity is determined by balancing the water viscous stress with the osmotic pressure gradient, resulting in (Siria, et al., 2013):

$$V_{\text{DO}} = \frac{k_B T}{8\pi l_B \eta}\frac{\Delta c}{cL} \tag{9}$$

where L and η are the length of the nanochannel and the viscosity of the solvent, respectively; and l_B is Bjerrum's length that is about 0.7 nm. Then the DO electric current is expressed as $I_{DO} \approx 2\pi R\Sigma V_{DO}$ for a nanochannel with high surface charges and $I_{DO} \approx \Sigma^3$ for a nanochannel with low charges, where Σ represents the net surface charge. When an external electric circuit is used to collect the generated electricity, the maximum power can be obtained:

$$P_{DO,max} = \frac{R_M I_{DO}^2}{4} \tag{10}$$

where R_M is the internal ionic resistance of the nanochannel.

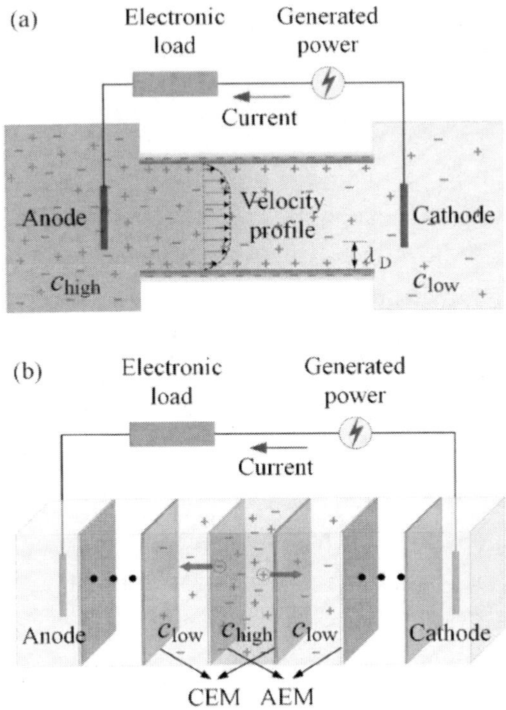

Figure 3. (a) Schematic of using Diffusio-osmosis (DO) based method for harvesting salinity gradient energy. DO is referred to as an electrokinetic phenomenon happening within a few nanometers region of a charged nanochannel wall. The DO energy conversion process generates a plug-like fluid flow which carries the excess free charges within the electric double layer (EDL). Figure adapted from Ref. (Siria, et al., 2013). (b) Schematic of Reverse electrodialysis (RED) energy conversion for harvesting salinity gradient energy. A stack of ion-selective membranes, including both anion-exchange membrane (AEM) and cation-exchange membrane (CEM) are alternately placed in between solutions with high concentration and low concentration alternately arranged. The salinity gradient, i.e., the osmotic pressure gradient, drives the ions across the nanochannel in DS and the CEM or the AEM in RED.

Figure 3 (b) shows the schematic of a reverse electrodialysis (RED) energy conversion for harvesting the salinity gradient energy. Specifically, a stack of ion-selective membranes, including both anion-exchange membrane (AEM) and cation-exchange membrane (CEM) are alternately placed in between solutions with high concentration and low concentration alternately arranged. The presence of salinity gradient between the low and high concentration solutions drives the ions across the CEM or the AEM in RED. Clearly, DO and RED have similar features that the nanochannels with surface charges for DO (the ion-selective membranes for RED) and the salinity gradient are needed for ion transportation in both DO and RED processes. A review study on the new avenues for the large-scale harvesting of blue energy considers that DO, unlike RED which requires selective membranes, allows to use nanopores with relatively large radii of several tens of nanometers (Siria, Bocquet, & Bocquet, 2017). However, the selectivity of the nanochannel highly depends on the relationship between the thickness of the electric double layer (EDL) λ and the nanochannel size. Specifically, the nanochannel exhibits selective when λ is larger than half of the nanochannel diameter or height. Since λ is a function of solution concentration, i.e., $\lambda = (\varepsilon_0\varepsilon_r\mathrm{RT}/2cF^2)^{1/2}$, the selectivity of the nanochannel eventually depends on the solution concentration which usually varies along the channel length direction. Thus, the nanochannel could be selective near to the low concentration side and nonselective near to the high concentration side (Feng et al., 2016). Practically, RED is frequently applied to the membranes with irregularly structured nanochannels, while DS is frequently applied to the well-structured/organized nanochannels or membranes.

4. Comparison and Outlook

Among the three osmosis based green power generation methods, the FO-EK and the PRO methods involve two modules: one is the development of the hydrodynamic force from salinity gradient energy through a module including a semipermeable membrane, and the other is the power generation module. Differently, the DO method directly converts the salinity gradient energy into electricity without need of any intermediate module. A study compared the PRO method with the RED method by considering the generated power density and the energy conversion efficiency, and found that the PRO method is superior to the RED method, particularly when extracting large salinity gradient energy from the solutions of relatively high concentrations (N. Y. Yip & Elimelech, 2014). In addition, all these three membrane-based energy conversion processes have great potential to be coupled with desalination processes, such as reverse osmosis (RO) (Feinberg, Ramon, & Hoek, 2013; Kim, Kim, Shon, & Hong, 2015; Kwon, Han, Park, Shin, & Kim, 2015; Prante, Ruskowitz, Childress, & Achilli, 2014) and membrane distillation (MD) (Han, 2015; Tufa et al., 2015). Through the coupling, it allows for compensation to the energy consumption and also recovery of the salinity energy from the concentrated brines, thereby effectively reducing desalination costs, and importantly mitigating the environmental impact.

The FO-EK power generation method was developed recently, and its power density has been achieved only several watts per square meter (membrane). Efforts need be made to improve its energy conversion efficiency through exploring new membranes and porous media consisting of micro/nanostructures (channels). Researches on the PRO power generation method showed that its power densities ranging from 1 W/m^2 to 5.1 W/m^2 were successfully obtained with the DS from both the seawater with relatively low concentration and the brines with relatively high concentrations discharged from desalination processes (Achilli, Cath, & Childress, 2009; N. Y. Yip, Tiraferri, Phillip, Schiffman, & Elimelech, 2010). A thin-film composite polyamide membrane was developed as the new semipermeable membrane used in the PRO process, and the resultant power density reached up to 5.7~10 W/m^2 (Chou et al., 2012; Ngai Yin Yip et al., 2011; N. Y. E. M. Yip, 2011). A pilot PRO plant, constructed by Statkraft Company, only yielded a power density of about 1 W/m^2 by using the commercially available asymmetrical cellulose-acetate membranes (Skilhagen, Dugstad, & Aaberg, 2008). Certainly, such power density is far below the expected power output of 5 W/m^2 that is required for commercialization. This is attributed to lack of effective membranes with minimum concentration polarization, maximum salt rejection, and maximum water permeation (Achilli, et al., 2009; Jellinek & Masuda, 1981;

S. Loeb & Mehta, 1978; Mehta & Loeb, 1978, 1979). Consequently, it has hindered the scale-up of the PRO power generation method.

In addition to the PRO method, the RED method for power generation from salinity gradient energy has been extensively studied. The earliest study reported a power density of about 0.2 W/m^2 only (Pattle, 1954). A pilot RED power plant equipped with 50 m^2 ion exchange membranes was tested for five months in the south of Italy, and the obtained power density was around 1.5~1.7 W/m^2 (Tedesco et al., 2016). More recently, another pilot RED power plant was built in Afsluitdijk, a dyke separating the Ijssel Lake from the Wadden Sea, and it was designed by using the salinity gradient between seawater and river water to generate electricity. It is expected to yield a power output of 0.5~2 MW and subsequently to be commercialized as a stand-alone power plant by 2020 (Li, Siddiqi, Anadon, & Narayanamurti, 2018). Additionally, instead of using conventional membranes, researchers have developed well-structured, nano-scale membranes/channels, and attained dramatically improvement in energy conversion from the salinity gradient energy. Siria. et al., investigated the DO phenomenon by using a hierarchical nanofluidic device made of boron nitride nanotubes (BNNTs) that are pierced into an ultrathin membrane and connected with two fluid reservoirs. For such BNNTs with their radii of 40 nm and length of 1.25 μm, a maximum power per unit membrane cross-section area of 4×10^3 W/m^2 was obtained based on the solution concentrations of c_{high} and c_{low} as 1 M and 10^{-3} M, respectively (Siria, et al., 2013). Of note, this power density level attained by the DO method is of about three orders higher than those from the RED, FO-EK and PRO methods. Feng, et al., investigated the DO process by using a thin MoS_2 membrane with its thickness of 0.65 nm, homogeneous pore size of 10 nm and porosity of 30%, and they achieved an impressive power density of 10^6 W/m^2 for two KCl solutions having their concentrations of 1 M and 10^{-3} M. This study also found that increase of the thickness of MoS_2 membranes would reduce in the mobility of ions and thus the generated power, suggesting that ultra-thin membranes can give rise to the best energy conversion performance for harvesting the salinity gradient energy (Feng, et al., 2016). Walker et al., applied an ozone-treated graphene in the DO energy conversion process and showed that a power density of 0.7×10^3 W/m^2 can be achieved with KCl solutions of pH 7 in two reservoirs at their concentrations of 1 M and 10^{-3} M. Most importantly, this study demonstrated that the enhancement in cation transport across the monolayer graphene membrane is resulted from the ozone treatment that allows to create more nanopores in the membrane while maintaining the selectivity of the original nanopores (Walker et al., 2017).

Based on the above-reviewed research works, their results have amply showed that the salinity energy conversion performance of the FO-EK, PRO, and RED methods has relatively low efficiency, while the DO method has relatively high efficiency. Moreover, their results also suggested that the most promising way of enhancing the energy conversion efficiency is to explore advanced membranes with tailored surface chemistry and highly ordered nanochannels. This will facilitate commercial applications of the osmosis-based salinity energy conversion methods such as the DO method. Therefore, the following three membrane parameters are expected to play important roles in the DO based the energy conversion process:

1) Surface charge density. This is crucial for efficient osmotic energy conversion. It is suggested to seek and use materials featured with more active ionic bonds, such as BN, MoS_2, and TiO_2. The high surface charge density not only depends on the active ionic bonds, but also on the charging mechanisms of the nanopore surface. Thus the pH strength is a key factor affecting the surface charges. It was found that non-monotonic increment of the energy conversion performance with increasing the surface charge density results from the pH-dependent ion concentration polarization effect, which causes the degradation in effective concentration ratio across the nanopore (Yeh, et al., 2017). In addition, chemical coating is another efficient way to augment the surface charges.
2) Porosity. Since the DO energy conversion process is associated to the effective passage of nanochannels, a high performance membrane would have highly ordered nanochannels or high porosity.

3) Thickness. Given the same concentration difference between the draw and feed solutions, a thinner membrane yields a larger osmotic pressure gradient, leading to a stronger driving force to transport the ions inside the membrane channels. Therefore, the membrane thickness is also a dominating factor in the DO process.

In addition to these three parameters, the membrane wettability is another factor affecting salinity gradient energy conversion processes. Xie, et al., studied the hydraulic resistance and slippage of single graphene nanochannels, and they found that the graphene nanofluidic channels can dramatically increase the slipping of water due to the presence of low surface charge density. (Xie et al., 2018). However, the low surface charge density may lead to high ion concentration polarization, thereby lowering down the performance of salinity energy conversion.

CONCLUSION

This chapter reviews the green power generation from osmosis based salinity gradient energy, with the main focus on the FO-EK, PRO, and DO three methods. A brief introduction is given on the fundamentals of osmosis. Subsequently, the working principles, the development as well as the energy conversion performance of these three methods are provided. Furthermore, a summary of the existing research works on the FO-EK, PRO, and DO methods suggest that: (1) The DO power generation method shows a great potential; (2) The well-structured nanochannels or membranes are the prerequisite factors for enhancing the energy conversion efficiency of the osmosis-related power generation methods; (3) Factors dominating the performance of the membranes used in the DO process include the surface charge density, porosity, thickness, as well as the surface wettability. More advanced membranes should be explored to further improve the DO energy conversion efficiency to a certain level needed for commercialization at industrial scale.

REFERENCES

Achilli, A., Cath, T. Y., & Childress, A. E. (2009). Power generation with pressure retarded osmosis: An experimental and theoretical investigation. *Journal of Membrane Science, 343*(1-2), 42-52.

Anderson, J. L. (1989). Colloid transport by interfacial forces. *Annual Review of Fluid Mechanics, 21*(1), 61-99.

Bouzigues, C. I., Tabeling, P., & Bocquet, L. (2008). Nanofluidics in the debye layer at hydrophilic and hydrophobic surfaces. *Physical Review Letters, 101*(11), 114503-n/a.

Brogioli, D. (2009). Extracting renewable energy from a salinity difference using a capacitor. *Physical Review Letters, 103*(5), 058501-n/a.

Chang, C. C., & Yang, R. J. (2009). Electrokinetic energy conversion in micrometer-length nanofluidic channels. *Microfluidics and Nanofluidics, 9*(2-3), 225-241.

Chein, R., Liao, C., & Chen, H. (2009). Electrokinetic energy conversion efficiency analysis using nanoscale finite-length surface-charged capillaries. *Journal of Power Sources, 187*(2), 461-470.

Chou, S., Wang, R., Shi, L., She, Q., Tang, C., & Fane, A. G. (2012). Thin-film composite hollow fiber membranes for pressure retarded osmosis (PRO) process with high power density. *Journal of Membrane Science, 389*, 25-33.

Daiguji, H., Yang, P. D., Szeri, A. J., & Majumdar, A. (2004). Electrochemomechanical energy conversion in nanofluidic channels. *Nano Letters, 4*(12), 2315-2321.

Eijkel, J. (2007). Liquid slip in micro- and nanofluidics: recent research and its possible implications. *Lab on a Chip, 7*(3), 299-301.

Einzel, D., Panzer, P., & Liu, M. (1990). Boundary condition for fluid flow: Curved or rough surfaces. *Physical Review Letters, 64*(19), 2269-2272.

Everett, D. H. (1988). *Basic principles of colloid science*. London: Royal Society of Chemistry.

Feinberg, B. J., Ramon, G. Z., & Hoek, E. M. V. (2013). Thermodynamic analysis of osmotic energy recovery at a reverse osmosis desalination plant. *Environmental Science & Technology, 47*(6), 2982-2989.

Feng, J., Graf, M., Liu, K., Ovchinnikov, D., Dumcenco, D., Heiranian, M., et al., (2016). Single-layer MoS2 nanopores as nanopower generators. *Nature, 536*, 197-200.

Han, G. Z. J. W. C. C. T. S. (2015). Hybrid pressure retarded osmosis-membrane distillation (PRO-MD) process for osmotic power and clean water generation. *Environmental Science: Water Research & Technology, 1*(4), 507-507.

Hon, K. C., Zhao, C., Yang, C., & Seow, C. L. (2012). A method of producing electrokinetic power through forward osmosis. *Applied Physics Letters, 101*(14), 143902-n/a.

Hunter, R. J. (1981). *Zeta potential in colloid science: Principles and applications*. London; New York: Academic Press.

Jellinek, H. H. G., & Masuda, H. (1981). Osmo-power - theory and performance of an osmo-power pilot-plant. *Ocean Engineering, 8*(2), 103-128.

Jiao, Y., Kang, Y., & Yang, C. (2015). Osmosis and its applications. In D. Li (Ed.), *Encyclopedia of Microfluidics and Nanofluidics* (pp. 2622-2633). New York, NY: Springer New York.

Jiao, Y., Yang, C., & Kang, Y. (2014). Energy conversion from salinity gradients by forward osmosis–electrokinetics. *The Journal of Physical Chemistry C, 118*(20), 10574-10583.

Jiao, Y., Yang, C., Kang, Y., & Hon, K. C. (2014). *Enhancement of elecrtrokinetic power generation by surface treatment on a porous glass*. Paper presented at the The 3rd International Conference on Informatics, Environment, Energy and Applications (IEEA 2014).

Joly, L., Ybert, C., Trizac, E., & Bocquet, L. (2006). Liquid friction on charged surfaces: from hydrodynamic slippage to electrokinetics. *The Journal of Chemical Physics, 125*(20), 204716-n/a.

Joly, L., Ybert, C., Trizac, E., Bocquet, L., eacute, & ric. (2004). Hydrodynamics within the electric double layer on slipping surfaces. *Physical Review Letters, 93*(25), 257805-n/a.

Kim, D. I., Kim, J., Shon, H. K., & Hong, S. (2015). Pressure retarded osmosis (PRO) for integrating seawater desalination and wastewater reclamation: Energy consumption and fouling. *Journal of Membrane Science, 483*, 34-41.

Kwon, K., Han, J., Park, B. H., Shin, Y., & Kim, D. (2015). Brine recovery using reverse electrodialysis in membrane-based desalination processes. *Desalination, 362*, 1-10.

La Mantia, F., Pasta, M., Deshazer, H. D., Logan, B. E., & Cui, Y. (2011). Batteries for efficient energy extraction from a water salinity difference. *Nano Letters, 11*(4), 1810-1813.

Lee, K. L., Baker, R. W., & Lonsdale, H. K. (1981). Membranes for power-generation by pressure-retarded osmosis. *Journal of Membrane Science, 8*(2), 141-171.

Lee, S., Boo, C., Elimelech, M., & Hong, S. (2010). Comparison of fouling behavior in forward osmosis (FO) and reverse osmosis (RO). *Journal of Membrane Science, 365*(1–2), 34-39.

Li, Z., Siddiqi, A., Anadon, L. D., & Narayanamurti, V. (2018). Towards sustainability in water-energy nexus: Ocean energy for seawater desalination. *Renewable and Sustainable Energy Reviews, 82*, 3833-3847.

Loeb, S. (1976). Production of energy from concentrated brines by pressure-retarded osmosis: I. Preliminary technical and economic correlations. *Journal of Membrane Science, 1*, 49-63.

Loeb, S., & Mehta, G. D. (1978). A two-coefficient water transport equation for pressure-retarded osmosis. *Journal of Membrane Science, 4*, 351-362.

Loeb, S., Van Hessen, F., & Shahaf, D. (1976). Production of energy from concentrated brines by pressure-retarded osmosis: II. Experimental results and projected energy costs. *Journal of Membrane Science, 1*(0), 249-269.

Mehta, G. D., & Loeb, S. (1978). Internal polarization in the porous substructure of a semipermeable membrane under pressure-retarded osmosis. *Journal of Membrane Science, 4*(2), 261-265.

Mehta, G. D., & Loeb, S. (1979). Performance of permasep B-9 and B-10 membranes in various osmotic regions and at high osmotic pressures. *Journal of Membrane Science, 4*(3), 335-349.

Olsson, M., Wick, G. L., & Isaacs, J. D. (1979). Salinity gradient power: Utilizing vapor pressure differences. *Science, 206*(4417), 452-454.

Pattle, R. E. (1954). Production of electric power by mixing fresh and salt water in the hydroelectric pile. *Nature, 174*, 660.

Prante, J. L., Ruskowitz, J. A., Childress, A. E., & Achilli, A. (2014). RO-PRO desalination: An integrated low-energy approach to seawater desalination. *Applied Energy, 120*, 104-114.

Prieve, D. C., Anderson, J. L., Ebel, J. P., & Lowell, M. E. (1984). Motion of a particle generated by chemical gradients. Part 2. Electrolytes. *Journal of Fluid Mechanics, 148*, 247-269.

Ramon, G. Z., Feinberg, B. J., & Hoek, E. M. V. (2011). Membrane-based production of salinity-gradient power. *Energy & Environmental Science, 4*, 4423-4434.

Rankin, D. J., & Huang, D. M. (2016). The effect of hydrodynamic slip on membrane-based salinity-gradient-driven energy harvesting. *Langmuir, 32*(14), 3420-3432.

Siria, A., Bocquet, M.-L., & Bocquet, L. (2017). New avenues for the large-scale harvesting of blue energy. *Nature Reviews Chemistry, 1*, 0091-n/a.

Siria, A., Poncharal, P., Biance, A.-L., Fulcrand, R., Blase, X., Purcell, S. T., et al., (2013). Giant osmotic energy conversion measured in a single transmembrane boron nitride nanotube. *Nature, 494*, 455-458.

Skilhagen, S. E., Dugstad, J. E., & Aaberg, R. J. (2008). Osmotic power — power production based on the osmotic pressure difference between waters with varying salt gradients. *Desalination, 220*(1), 476-482.

Strathmann, H. (2004). *Ion-exchange membrane separation processes*. Burlington: Elsevier.

Tedesco, M., Scalici, C., Vaccari, D., Cipollina, A., Tamburini, A., & Micale, G. (2016). Performance of the first reverse electrodialysis pilot plant for power production from saline waters and concentrated brines. *Journal of Membrane Science, 500*, 33-45.

Tufa, R. A., Curcio, E., Brauns, E., van Baak, W., Fontananova, E., & Di Profio, G. (2015). Membrane distillation and reverse electrodialysis for near-zero liquid discharge and low energy seawater desalination. *Journal of Membrane Science, 496*, 325-333.

Walker, M. I., Ubych, K., Saraswat, V., Chalklen, E. A., Braeuninger-Weimer, P., Caneva, S., et al., (2017). Extrinsic Cation Selectivity of 2D Membranes. *ACS Nano, 11*(2), 1340-1346.

Xie, Q., Alibakhshi, M. A., Jiao, S. P., Xu, Z. P., Hempel, M., Kong, J., et al., (2018). Fast water transport in graphene nanofluidic channels. *Nature Nanotechnology, 13*(3), 238-245.

Yeh, L. H., Chen, F., Chiou, Y. T., & Su, Y. S. (2017). Anomalous pH-dependent nanofluidic salinity gradient power. *Small, 13*(48), 1702691-n/a.

Yip, N. Y., & Elimelech, M. (2014). Comparison of energy efficiency and power density in pressure retarded osmosis and reverse electrodialysis. *Environmental Science & Technology, 48*(18), 11002-11012.

Yip, N. Y., Tiraferri, A., Phillip, W. A., Schiffman, J. D., & Elimelech, M. (2010). High performance thin-film composite forward osmosis membrane. *Environmental Science & Technology, 44*(10), 3812-3818.

Yip, N. Y., Tiraferri, A., Phillip, W. A., Schiffman, J. D., Hoover, L. A., Kim, Y. C., et al., (2011). Thin-film composite pressure retarded osmosis membranes for sustainable power generation from salinity gradients. *Environmental Science & Technology, 45*(10), 4360-4369.

Yip, N. Y. E. M. (2011). Performance limiting effects in power generation from salinity gradients by pressure retarded osmosis. *Environmental Science & Technology, 45*(23), 10273-11028.

Zhang, J., Loong, W. L. C., Chou, S., Tang, C., Wang, R., & Fane, A. G. (2012). Membrane biofouling and scaling in forward osmosis membrane bioreactor. *Journal of Membrane Science, 403*, 8-14.

CONTENTS OF EARLIER VOLUMES

Advances in Energy Research. Volume 29

Advances in Energy Research. Volume 28

INDEX

A

B

C

D

E

S

T

V

W